THE SCIENCE
OF MARIJUANA

THE SCIENCE
OF MARIJUANA

THIRD EDITION

Leslie L. Iversen

OXFORD
UNIVERSITY PRESS

OXFORD
UNIVERSITY PRESS

Oxford University Press is a department of the University of Oxford. It furthers
the University's objective of excellence in research, scholarship, and education
by publishing worldwide. Oxford is a registered trade mark of Oxford University
Press in the UK and certain other countries.

Published in the United States of America by Oxford University Press
198 Madison Avenue, New York, NY 10016, United States of America.

Library of Congress Cataloging-in-Publication Data
Names: Iversen, Leslie L., author.
Title: The science of marijuana / Les Iversen.
Description: 3rd Edition. | New York : Oxford University Press, [2018] |
Revised edition of the author's The science of marijuana, 2008.
Identifiers: LCCN 2018016007 | ISBN 9780190846848
Subjects: LCSH: Marijuana—Physiological effect. | Marijuana—Toxicology.
Classification: LCC QP801.C27 I94 2018 | DDC 615.9/523648—dc23
LC record available at https://lccn.loc.gov/2018016007

1 3 5 7 9 8 6 4 2

Printed by Sheridan Books, Inc., United States of America

CONTENTS

PREFACE

Marijuana—known as cannabis in Europe—has been used medically and recreationally for thousands of years. There is a considerable scientific literature on cannabis and the possible beneficial or adverse effects associated with its use. Some societies have regarded cannabis as a sacred boon for mankind that offers respite from the tribulations of everyday life, whereas others have demonized it as inevitably leading to "reefer madness." Millions of young people on both sides of the Atlantic are more or less regular users of cannabis, but official attitudes vary widely. In Europe, several countries have relaxed the legal penalties associated with its use. In the United States, several states have sanctioned the medical use of cannabis, and some states have even approved its recreational use; however, these states remain in conflict with the federal government, which continues to view cannabis as an illegal and harmful drug. Cannabis accounted for 643,00 arrests in 2016 in the United States, which were often followed by draconian penalties.

Research on the naturally occurring cannabinoids ("endo-cannabinoids") has grown rapidly since the first edition of this book was published in 2001. Less than 200 scientific papers had been published on these newly discovered chemicals by the time of the first edition. More than 2,000 additional publications appeared by the time of the second edition in 2008, and since then the number of citations in PubMed for endocannabinoids has risen to 5,384, that for THC has risen from 800 to 8,700, and the total number of citations for cannabis in PubMed stands at 16,180. This is clearly an active field of research, and there is an increasing understanding that the naturally occurring cannabinoid system plays many roles in the body apart from acting as modulators of neural activity in the brain. The potential for genuinely new medicines based on our knowledge of the cannabinoids is huge and is only now being exploited:

> There is likely no greater hot-button topic in biomedical and clinical research today than discussions surrounding the utility of active chemicals found in marijuana and how they interact within the body's endocannabinoid system to produce beneficial or deleterious effects. (Gordon Research Conference, "Cannabinoid Function in the CNS," 2017; https://www.grc.org/cannabinoid-function-in-the-cns-conference/2017)

Leslie L. Iversen
Professor
Department of Pharmacology
University of Oxford
Oxford, UK

Chapter 1

History and the Plant

A Brief History

Reviews of the history of cannabis can be found in Abel (1943), Lewin (1931), Walton (1938), Robinson (1996), Russo (2016), and Ligresti et al. (2016).

Evidence of the first use of cannabis by humans has been found in fragments of pottery bearing the imprint of a cord-like material thought to be of hemp in Taiwan, dated approximately 10,000 BC. Other early evidence for hemp cultivation derives from the finding of fragments of hemp cloth in Chinese burial chambers from the Chou dynasty (1122–265 BC). Hemp is a strain of *Cannabis sativa* containing little or no psychoactive ingredients, cultivated for its tough fibers. It seems likely that hemp was cultivated and used for the manufacture of ropes, nets, canvas sails, and cloths in ancient China. The first descriptions of the medical and intoxicant properties of the plant are to be found in the ancient Chinese herbal, *Pen-ts'ao*, ca. 1–2 century AD. Classical myth relates that the Chinese deity Shen Nung tested hundreds of herbal materials in a series of heroic experiments in self-medication and agronomics. So potent was this myth of the etiology of medicine that the god's name was attached to the *Pen-ts'ao*. This herbal pharmacopeia describes hundreds of drugs, among them cannabis, which was called *ma*, a pun for "chaotic." This ancient text clearly describes the stupefying and hallucinogenic properties of the plant. Pharmacologists and herbalists added sections to the text for many centuries, and Chinese physicians used cumulative editions of *Pen-ts'ao* as the standard text on medical drugs for hundreds of years. Shen Nung, the Farmer God, became the patron deity of medicine, with the title "Father of Chinese Medicine." Ma, often mixed with wine in a preparation called *ma-yao*, was used principally for its pain-relieving properties. Although there seems also to have been some use of the drug as an intoxicant in China, this never became widespread.

In contrast to China, the use of cannabis for its psychoactive properties was known by the nomadic tribes of northeastern Asia in Neolithic times, and it may have played an important role in the practice of the religion of shamanism by these people. The nomads brought the plant and its uses to Western Asia and then to India. Ancient Indian legend tells how the Hindu god Siva became angry after a family row and wandered off into the fields by himself. Exhausted by the heat of the sun, he sought shade and refuge under a leafy plant and finally went to sleep. On waking, he became curious about the plant that had given him shelter and ate some of its leaves. This made him feel so refreshed that he adopted it as his favorite food. From then on, Siva was known as the Lord of bhang. In ancient Indian texts, bhang is referred to in the *Science of Charms*—written between 2000 and 1400 BC—as one of the "five kingdoms of herbs . . . which release us from anxiety." Bhang seems to have been popular with the Indian people from the beginning of history. The Indian Hemp Drugs Commission Report (1894) gave a detailed picture of how bhang and the more potent cannabis products ganja and charas (the Indian term for cannabis resin) had become incorporated into Indian life and culture.

It took longer for cannabis to reach the West. Hemp was known to the Assyrian civilization both as a fiber plant and as a medicine, and it is referred to as *kunnubu* or *kunnapu* in Assyrian documents of approximately 600 BC. The word is probably the basis of the Arabian *kinnab* and the Greek and Latin *cannabis*. There is little evidence that the plant was known beyond Turkey until the time of the Greeks. The Greeks used hemp for the manufacture of ropes and sails for their conquering navies, as did the Romans later—although hemp was not cultivated in Greece or Italy but, rather, in the further reaches of their empires in Asia Minor. Neither the Greeks nor the Romans, however, appear to have used cannabis for its psychoactive properties, although these were known and described by the Roman physicians Dioscorides, Galen, and Oribasius. Galen, writing in the 2nd century AD, described how wealthy Romans sometimes offered their dinner guests an exotic dessert containing cannabis seeds:

> There are those who eat it (cannabis seed) also cooked with other confections, by this confection is meant a sort of dessert which is taken after meals with drinks for the purpose of exciting pleasure. It creates much warmth (or possibly excitement) and when taken too generously affects the head emitting a warm vapor and acting as a drug. (Walton, 1938, p. 8)

Because the seeds contain no significant amounts of psychoactive material, it seems likely that some other parts of the cannabis plant must also have been included.

It was to be almost another 1,000 years before cannabis spread to the Arab lands and then to Europe and the Americas. According to one Arab legend, the discovery of marijuana dates back to the 12th century AD when a monk and recluse named Hayder, a Persian founder of the religious order of Sufi, came across the plant while wandering in meditation in the mountains. When he returned to his monastery after eating some cannabis leaves, his disciples were amazed at how talkative and animated this normally dour and taciturn man had become. After they persuaded Hayder to tell them what had made him so happy, the disciples went out into the mountains and tried some cannabis themselves. By the 13th century, cannabis use had become common in the Arab lands, giving rise to many colorful legends. Bhang and hashish are referred to frequently in the *Arabian Nights* or *The Thousand and One Nights* folk tales collected during the period 1000 to 1700 AD:

> Furthermore, I conceive that the twain are eaters of Hashish, which
> drug when swallowed by man, maketh him prattle of whatso he
> pleaseth and chooseth, making him now a Sultan, then a Wazir,
> and then a merchant, the while it seemeth to him that the world is
> in the hollow of his hand. Tis composed of hemp leaflets whereto
> are added aromatic roots and somewhat of sugar; then they cook it
> and prepare a kind of confection which they eat, but whoso eateth
> it (especially if he eat more than enough) talketh of matters which
> reason may on no wise represent. (Walton, 1938, p. 15)

It is clear from this description that the word "hashish" in ancient
Arab writings refers to what we would now call "herbal marijuana"
rather than the cannabis resin, which the term *hashish* now describes.
Outstanding among the Arab legends is the story of the "Old Man
of the Mountains" and his murderous band of followers known as
the "Assassins." According to Marco Polo, who recorded this legend,
the Assassins were led by the "Old Man of the Mountains," who
recruited novices to his band and kept them under his control as
his docile servants by feeding them copious amounts of hashish.
Marco Polo described how the leader constructed a remarkable
garden at his major fortress, the Alamut. The young Assassins would
be transported to the garden after they had taken enough hashish to
put them to sleep. When they awoke and found themselves in such a
beautiful place with ladies willing to dally with them to their heart's
content, they believed that they were indeed in paradise. When the
Old Man wanted someone killed, he would tell the Assassins to do it
and promise them that, dead or alive, they would return to paradise;
they obeyed his commands with great brutality.

Although the historical facts are impossible to determine, it
seems likely that the Assassins were led by Hasan-Ibn-Sabbah, who
started life as a religious missionary and later gathered a secret band

of followers. They probably used hashish, as did many others in the Arab world at that time. It does not seem likely that they would have been able to carry out their terrorist acts or politically motivated assassinations while intoxicated by cannabis, nor is there any significant evidence that the drug inspires violence—on the contrary, it tends to cause somnolence and lethargy when taken in high doses. Nevertheless, lurid stories about the drug-crazed Assassins have been widely used in the West as part of the mythology that surrounds the cannabis debate. As early as the 12th century, Abbot Arnold of Lübeck wrote in *Chronica Slavorum* (*Abbot Arnold of Lubeck, 1868 Chronica Slavorum, Hannnoverae, Impensus, Biblopoli Hahniani.*):

> Hemp raises them to a state of ecstasy or folly, or intoxicates them. Then sorcerers draw near and exhibit to the sleepers phantasms, pleasures and amusements. They then promise that these delights will become perpetual if the orders given them are executed with the daggers provided.

Eight hundred years later in the United States, the hard line Commissioner of the Federal Bureau of Narcotics, Harry J. Anslinger, used the image of the drug-crazed assassin in his personal vendetta against the drug. He wrote in *American Magazine* in 1937:

> In the year 1090, there was founded in Persia the religious and military order of the Assassins, whose history is one of cruelty, barbarity and murder, and for good reason. The members were confirmed users of hashish, or marijuana, and it is from the Arab "hashishin" that we have the English word "assassin." (Anslinger and Cooper, 1937, p. 150)

The use of cannabis was particularly common in Egypt during the Middle Ages, where the "Gardens of Cafour" in Cairo became a notorious haunt of hashish smokers. Despite draconian measures by the Egyptian authorities to close such establishments and to prohibit hashish use during the thirteenth and fourteenth centuries, the habit had become too firmly ingrained in the Arab world for it to be stamped out. The social acceptance of cannabis use among the people of Egypt and other Arab lands was reinforced by the fact that although the holy Koran explicitly banned the consumption of alcohol, it did not mention cannabis. Not all were happy about this acceptance of cannabis, however. Ebn-Beitar wrote of the spread of cannabis use in Egypt 600 years ago:

> It spread insensibly for several years and became of common enough usage that in the year 1413 A.D., this wretched drug appeared publicly, it was eaten flagrantly and without furtiveness, it triumphed.... One had no shame in speaking of it openly.... Also as a consequence of that, baseness of sentiments and manners became general; shame and modesty disappeared among men, they no longer blushed to hold discourse on the most indecent things. . . . And they came to the point of glorifying vices. All sentiments of nobility and virtue were lost. . . . And all manner of vices and base inclination were displayed openly. (Walton, 1938, p. 14)

It was from Egypt that the use of cannabis as a psychoactive drug first spread to Europe and then to the Americas. When Napoleon invaded and conquered Egypt at the end of the 18th century, he was dismayed by what he saw as the corrupting influence of hashish on the local population and the possible debilitating effects it might have on his own soldiers, who soon developed a liking for cannabis in this wine-free country. In 1800, one of his generals issued a decree:

> Article 1: Throughout Egypt the use of a beverage prepared by some Moslems from hemp (hashish) as well as the smoking of the seeds of hemp, is prohibited. Habitual smokers and drinkers of this plant lose their reason and suffer from violent delirium in which they are liable to commit excesses of all kinds.

Article 2: The preparation of hashish as a beverage is prohibited throughout Egypt. The doors of those cafes and restaurants where it is supplied are to be walled up, and their proprietors imprisoned for three months.

Article 3: All bales of hashish arriving at the customs shall be confiscated and burnt. (Lewin, 1931)

As with all the earlier bans, the Egyptians too largely ignored this one and Napoleon's army was soon to leave in retreat. However, the returning French army brought back to Europe many colorful tales of hashish and its intoxicating effects. Although cannabis had been cultivated in Europe for many centuries as a source of rope, canvas, and other cloths and in making paper, its inebriating effects were largely unknown—although secretly some sorcerers and witches may have included cannabis in their mysterious concoctions of drugs. In the mid-19th century in France, it became fashionable among a group of writers, poets, and artists in Paris's Latin Quarter to experiment with hashish. Among these was the young French author Pierre Gautier, who became so enthused by the drug that he founded the famous "Club des Hashischins" in Paris and introduced many others among the French literary world to its use. These included Alexander Dumas, Gerard de Nerval, and Victor Hugo—all of who wrote about their experiences with hashish. Gautier and his sophisticated literary colleagues regarded cannabis as an escape from a bourgeois environment, and they described their drug-induced experiences in flowery, romantic language. Thus, Gautier wrote the following:

> After several minutes a sense of numbness overwhelmed me. It seemed that my body had dissolved and become transparent. I saw very clearly inside me the Hashish I had eaten, in the form of an emerald which radiated millions of tiny sparks. The lashes of my eyes elongated themselves to Infinity, rolling like threads of gold on little ivory wheels, which spun about with an amazing rapidity. All around me I heard the shattering and crumbling of jewels of all

colours, songs renewed themselves without ceasing, as in the play of a kaleidoscope. (Walton, 1938, p. 59)

Among the most influential of Gautier's colleagues was Charles Baudelaire, whose book *Les Paradis Artificiels*, published in Paris in 1860, described the hashish experience in romantic and imaginative language:

> The senses become extraordinarily acute and fine. The eyes pierce Infinity. The ear perceives the most imperceptible in the midst of the sharpest noises. Hallucinations begin. External objects take on monstrous appearances and reveal themselves under forms hitherto unknown. They then become deformed and at last they enter into your being or rather you enter in to theirs. The most singular equivocations, the most inexplicable transpositions of ideas take place. Sounds have odour and colours are musical.

The book captured the imagination of many readers in the West and inspired further interest in the use of cannabis; it is still one of the most comprehensive and impressive accounts of the effects of cannabis on the human psyche. The use of hashish, however, did not become widespread in Europe. Cannabis use was practically unknown in Britain, for example, until the 1960s, although hemp had been cultivated for hundreds of years as a fiber and food crop. Similarly, in North America the hemp plant was imported soon after the first settlements and was widely cultivated. Kentucky was particularly renowned for its hemp fields, and Kentucky Hemp, selected for its fiber production, is an important fiber variety. Americans seemed unaware of the peculiar properties of cannabis, and it is also unlikely that the varieties selected for hemp fiber production contained significant amounts of delta-9-tetrahydrocannabinol (THC). The canvas sails and hemp ropes used by the American and British navies required large amounts of hemp fiber. The British Admiralty was always keen to find new sources of supply. Ship captains were given supplies of hemp seed to stimulate production overseas, and British colonies were obliged to cultivate hemp.

It was not until the well-known mid-19th-century American author Bayard Taylor wrote a lurid account of his experiences with hashish in the Middle East that there was any awareness of the psychoactive effects of cannabis. Taylor described what happened after taking a large dose of the drug:

> The spirit (demon, shall I not rather say?) of Hasheesh had entire possession of me. I was cast upon the flood of his illusions, and drifted helplessly withersoever they might choose to bear me. The thrills which ran through my nervous system became more rapid and fierce, accompanied with sensations that steeped my whole being in inutterable rapture. I was encompassed in a seal of light, through which played the pure, harmonious colours that are born of light. . . . I inhaled the most delicious perfumes; and harmonies such as Beethoven may have heard in dreams but never wrote, floated around me. (Walton, 1938, p. 65)

Taylor's accounts were intentionally sensational and played to the 19th-century appetite for tales of adventure and vice in faraway places. It is unlikely that many readers were encouraged to experiment with cannabis themselves. One exception, however, was a young man named Fitz Hugh Ludlow. Ludlow experimented with many drugs, and he started taking cannabis, then widely available in the United States in various pharmaceutical preparations. Ludlow's detailed accounts of his experiences and his subsequent addiction to cannabis are described in his book, *The Hasheesh Eater* (1857). Ludlow was an intelligent youth aged 16 years when he discovered cannabis in the local drug store, where he had already experimented with ether, chloroform, and opium. He used cannabis intensely for the next 3 or 4 years and wrote of his experiences as part of his subsequent withdrawal from the drug. The book has become a classic in the cannabis literature, equivalent in importance to Baudelaire's *Les*

Paradis Artificiels, and it is referred to again in Chapter 3. Ludlow's book, however, seems to have had little impact at the time of its publication. One reviewer of his book, writing in 1857, commented that America was fortunately "in no danger of becoming a nation of hasheesh eaters."

For almost 100 years from the mid-19th century until 1937, cannabis enjoyed a brief vogue in Western medicine (see Chapter 5). Following its introduction from Indian folk medicine, first to Britain and then to the rest of Europe and to the United States, a variety of different medicinal cannabis products were used.

The cannabis plant was introduced to Latin America and the Caribbean as early as the first half of the 15th century by slaves brought from Africa. It became fairly widely used in many countries in this region for its psychoactive properties, both as a recreational drug and in connection with various native Indian religious rites (see Chapter 7). The term *marijuana,* a Spanish–Mexican word originally used to describe tobacco, came into general use to describe cannabis in both South and North America.

The history of marijuana use in the United States and its prohibition has been told many times (Abel, 1943; Snyder, 1971). After a brief vogue in the mid-19th century, the popularity of marijuana waned, and it was only regularly used in the United States in a few large cities by local groups of Mexicans and by African American jazz musicians. It was the wave of immigrants who entered the southern United States from Mexico in the early decades of the 20th century, bringing marijuana with them, that first brought the drug into prominence in America—and eventually led to its prohibition. It came initially to New Orleans and some other southern cities and spread slowly in some of the major cities. There were colorful accusations that marijuana use provoked violent crime and corrupted the young, causing "reefer madness." The head of the Federal Narcotics Bureau, Harry Anslinger, waged an impassioned campaign to outlaw the drug. He was the original "spin doctor" of his time, cleverly manipulating other government agencies, popular opinion, and the media with lurid tales of the supposed evils of cannabis. In 1937, the US Congress, almost by default, passed the Marijuana Tax Act that effectively banned any further use of the drug in medicine and outlawed it as a dangerous narcotic. Use of the drug continued to grow, however, and by the late 1930s newspapers in many large cities were filled with alarming stories about this new "killer drug."

In 1937, no less than 28 different pharmaceutical preparations were available to American physicians, ranging from pills, tablets, and syrups containing cannabis extracts to mixtures of cannabis with other drugs—including morphine, chloroform, and chloral. American pharmaceutical companies had begun to take an active interest in research on cannabis-based medicines. The hastily approved Marijuana Tax Act of 1937 put a stop to all further medical use and essentially terminated all research in the field for another 25–30 years. In Britain, as in many other European countries, cannabis continued to have a limited medicinal use for much longer, but this declined as more reliable new medicines became available. "Tincture of Cannabis" was finally removed from the *British Pharmacopoeia* in the early 1970s, as the growing recreational use of cannabis was made illegal in the Misuse of Drugs Act 1971.

The "demonization" of cannabis in the United States soon after its arrival from Latin America colored attitudes to the drug for more than 100 years—not only in North America but also worldwide.

Twenty-First-Century Revival of Cannabis

Marijuana (cannabis) is among the most widely used of all psychoactive drugs. Despite the fact that its possession and use is illegal in many countries, cannabis is used regularly by as many as 20 million people in the United States and Europe and by millions more in other areas of the world. Thousands of patients with AIDS, multiple sclerosis, and a variety of other disabling diseases use marijuana in the firm belief that it makes their symptoms better, despite the relative paucity of medical evidence to substantiate this claim. The 21st century has seen a growing acceptance of the medical use of marijuana and moves toward its complete legalization in several US states and also in Europe. Since the turn of the century, voters in 31 US states approved propositions making marijuana available for medical use with a doctor's recommendation, and in further states voters approved initiatives in the 2016 election to legalize its use (see Chapter 5). In Europe, several countries have permitted medical cannabis, and others are contemplating such legislation (Talking Drugs, 2017).

The Netherlands pioneered the separation of cannabis from "hard" drugs such as cocaine or heroin in the 1970s, and it established licensed "coffee shops" for the legal supply of small quantities of cannabis. Colorado and Washington were the first US states to approve the availability of recreational cannabis (see Chapter 7). In the US election in November 2016, and subsequently, Massachusetts, Nevada, and California joined Colorado, Washington, Oregon, Alaska, Oklahoma, Vermont and the District of Columbia by voting for initiatives that make it legal for adults to consume cannabis (see Chapter 7).

There are some indications that Western society is starting to take a more liberal view toward cannabis use—one that tends toward the Dutch assessment of it as a "soft" drug that should be distinguished and separated from "hard" drugs. However, fierce opposition to cannabis use remains in many quarters. The US federal government continues to view cannabis as a dangerous drug and imposes harsh penalties for possession or dealing. With Attorney General Sessions, who is fiercely opposed to marijuana, enforcement of the federal law may be even more actively employed. In Europe, reports that teenage cannabis use might lead to mental illness in later life have gained a great deal of prominence (see Chapter 6). Even in liberal Holland, the coffee shops had until recently no legal means of obtaining their supplies of cannabis, and the Dutch government is under considerable pressure from nearby

European countries to modify its policy. With the absence of customs borders in the European Union, it is very easy for people from neighboring France, Germany, or Belgium to stock up on cannabis from Dutch outlets. There are political moves to limit access to the coffee shops to Dutch nationals only.

Who is right? Is cannabis a relatively harmless "soft" drug? Does it have genuine medical uses that cannot be fulfilled by other medicines? Or is the campaign to legalize the medical use of cannabis merely a smokescreen used by those seeking the wider acceptance of the drug? Is cannabis in fact an addictive narcotic drug that governments are right to protect the public from? This book reviews the scientific and medical evidence on cannabis and tries to answer some of these questions. Often, in analyzing the mass of scientific data, it is difficult to come to clear-cut conclusions. To make matters worse in this particular case, the opposing factions in the cannabis debate often interpret the same scientific evidence differently to suit their own purposes.

In July 1996, the British Minister of Health, in reply to a Parliamentary question about the medical uses of cannabis, stated, "At present the evidence is inconclusive. The key point is that a cannabis-based medicine has not been scientifically demonstrated to be safe, efficacious and of suitable quality." In August of that year, General Barry McCaffrey, the US drug czar, somewhat more bluntly said, "There is not a shred of scientific evidence that shows that smoked marijuana is useful or needed. This is not medicine. This is a cruel hoax." But time has shown them both to be wrong. There have been important advances in the medical applications of cannabis in the past few years, with the first large-scale clinical trials of cannabis-based medicines and the approval of one such prescription medicine. Meanwhile, the new scientific knowledge of naturally occurring cannabinoids in the body has offered entirely new approaches to the discovery and development of novel cannabinoid-based medicines.

The Plant

The hemp plant (*C. sativa*) has been cultivated as a multipurpose economic plant for thousands of years, and through the process of selection for various desirable characteristics, many different cultivated varieties exist—some grown exclusively for their fiber content, others for their content of psychoactive chemicals. *Cannabis sativa* is a lush, fast-growing annual, which can reach maturity in 60 days when grown indoors under optimum heat and light conditions and in 3–5 months in outdoor cultivation. The plant has characteristic finely branched leaves subdivided into lance-shaped leaflets with a sawtooth edge. The woody, angular, hairy stem may reach a height of 15 feet or more under optimum conditions. A smaller, more bushy subspecies reaching only approximately 4 feet in height known as *Cannabis indica* was first described by Lamark and is recognized by some modern botanists as a distinct species. The cannabis plant is either male or female. The male plant produces an obvious flower head, which produces pollen, whereas the female flower heads are less obvious and ensheathed in green bracts and hairs (Figure 1.1).

More than 100 cannabinoid chemicals have been isolated from the plant (ElSouhly et al., 2014). The principal psychoactive chemical is THC (Figure 1.2), with minor amounts of equally active delta-8-tetrahydrocannabinol and the non-psychoactive substance cannabidiol (CBD), often present in similar amounts to THC. The inactive substance cannabinol is thought to represent an oxidation product formed during storage.

THC is present in most parts of the plant, including the leaves and flowers, but it is most highly concentrated in fine droplets of sticky resin produced by glands at the base of the fine hairs which coat the leaves and particularly the bracts of the female flower head ("buds"). If pollinated, the female flower head will develop seeds; these contain no THC but have a high nutritional value. Indeed, cannabis was an important food crop—listed as one

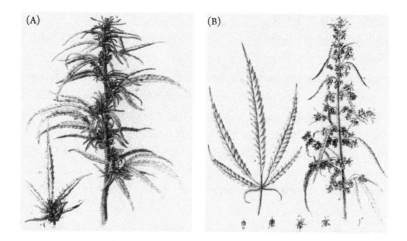

FIGURE 1.1 Engravings showing the characteristic appearance of the flowering heads of female (A) and male (B) cannabis plants.

Source: From Wisset (1808).

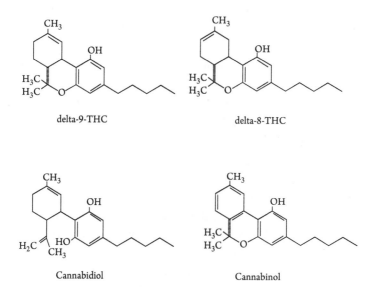

delta-9-THC

delta-8-THC

Cannabidiol

Cannabinol

FIGURE 1.2 Naturally occurring cannabinoids in cannabis extracts; delta-9-THC is the main psychoactive ingredient.

of the five major "grains" in ancient China. From the viewpoint of the cannabis smoker, however, the presence of seeds is undesirable: They burn with an acrid smoke and tend to explode on heating, and their presence dilutes the THC content of the female flower head. In the cultivation of cannabis for drug use in India, it was customary to remove all the male plants from the crop as they began to flower in order to yield the resin-rich sterile female flowering heads, which were dried and compressed to form the potent product known as "ganja." The services of expert "ganja doctors" were often employed, who went through the hemp field with an expert eye cutting down all the male plants before they could flower. The labor-intensive process of removing all male plants is rarely used by Western growers today; female plants can be cultivated simply by taking cuttings or using seed genetically modified to produce female-only plants. Nowadays, culture of cannabis often takes place indoors, where nutrients, lighting, and temperature conditions can be optimized, and the cultivation, where it is illegal, can be more easily concealed. Legal and illegal cannabis "farms" have multiplied on both sides of the Atlantic. Plants are grown outdoors, or indoors on specially enriched soils or with hydroponics, and their growth cycle has been shortened to less than 4 months (Potter, 2014). Cultivated cannabis, sometimes known as "skunk" or "Niederweit," has a higher THC content than traditional imported cannabis. The increased availability of such artificially cultivated cannabis at the turn of the century led to concerns that this higher potency material may be more dangerous than old-fashioned marijuana. Government claims in the United States that the potency levels of cultivated marijuana were 10–20 higher than previously reported were misleading. In reality, cultivated cannabis with a THC content of approximately 20% is not more than 4 or 5 times more potent than traditional marijuana. Such warnings were widely quoted, however, and became firmly embedded as a "media myth," accepted for a while by reputable newspapers and by the broadcast media as an established truth.

Among cannabis users, *c.indica* strains are said to be physically sedating, perfect for relaxing with a movie or as a nightcap before bed. *C. Sativas* typically provide more invigorating, uplifting cerebral effects that pair well with physical activity, social gatherings, and creative projects. Hybrids tend to fall somewhere in between on the indica–sativa spectrum, depending on the traits they inherit from their parent strains. This may be only partly true, however, because the profile of cannabinoids in any plant may vary considerably under different growing environments, making the distinction between *C. sativa* and *C. indica* less clear-cut. Extensive breeding programs to produce new strains of cannabis involve such techniques as the treatment of cannabis seed with the chemical colchicine to cause the creation of polyploid plants, in which each cell contains multiple sets of chromosomes instead of the normal single set. Such varieties may have extra vigor and an enhanced production of THC, although they tend to be genetically unstable. However, favorable genetic strains can also be propagated vegetatively by cuttings; in this way, a single plant can give rise to thousands of "clones" with identical genetic make-up to the original. Other varieties have been obtained by crossing *C. sativa* with *C. indica* strains to yield a number of different hybrids that may then be crossed with each other.

The legalization of cannabis in Colorado has led to a proliferation of stores and online sites from which cannabis and related products can be purchased (Helping Hands Herbals, 2017; see Chapter 7). Given the bewildering number of varieties now available, what is one to make of the various claims for their distinct psychoactive effects? One hypothesis is that *C. sativa* and *C. indica* differ genetically in their expression of enzymes catalyzing the conversion of the precursor cannabigerol to THC or CBD. *Cannabis sativa* strains express more of the enzyme-synthesizing CBD and tend to have lower THC:CBD ratios compared to *C. indica* strains (Hillig and Mahlberg, 2004). However, the expression of such enzymes varies considerably according to the growing environment. Another important variable may be the profile of different terpenes expressed in different varieties. Terpenes are a large group of volatile unsaturated hydrocarbons found in the essential oils of many plants, especially conifers and citrus trees. In cannabis plants, they are secreted by the same glands that produce THC and CBD. Terpenes are the pungent oils that color cannabis varieties and impart distinctive flavors such as citrus, berry, mint, and pine. More than 100 different terpenes have been identified in the cannabis plant, and every strain of cannabis has a different profile. In addition to their aromas, there is evidence that terpenes may contribute to the psychoactive effects of different strains (Leafly, 2017).

Cannabis Preparations

The dried leaves and small stems of the plant represent "herbal cannabis," a low-potency preparation that is little used today. The dried female flower heads ("buds") are the most common form. A preparation derived directly from the plant is "resin" ("hash"), which represents the THC-rich cannabis resin obtained by scraping the resin from the flower heads or by rubbing the dried flower heads and leaves through a series of sieves to obtain the dried particles of resin known as "polm." These are compressed to form a block of material. The process not only reduces the space required for storage but also ensures longer storage life by reducing the potential deterioration of herbal material through rot, mold, or infestation. The solid block of resin becomes sealed in its own oxidized coating. Resin was common in Western European countries when they relied on the import of large amounts of cannabis from overseas countries such as Thailand or Morocco. Currently, however, most of the cannabis for medical or recreational use is home-grown in "cannabis farms." Another product is "cannabis oil," which is produced by repeatedly extracting resin with alcohol. It can contain up to an alarmingly high 40% THC content, but usually the THC content is approximately

TABLE 1.1 Cannabis Preparations

Name	Part of Plant	THC %
Herbal marijuana (cannabis)	Leaves and small stems	4–6
Buds (sinsemilla)	Female flowers	10–30
Resin (hash)	Compressed bracts	15–25
Cannabis oil	Alcohol extract	20–40
Concentrates (dubs)	CO_2 or butanol extract	40–80

20%. Recently, a number of even stronger preparations have become available as waxes, oils, or solids obtained by solvent (liquid CO_2 or butane) extraction, yielding products known as "concentrates" containing as much as 80% THC. One drop of cannabis oil or a tiny piece of other concentrates can contain as much THC as a single marijuana cigarette. The concentrations of THC in various preparations are summarized in Table 1.1.

The legalization of cannabis in some US states and in some European countries has removed the need for secrecy and has led to the possibility of cannabis farms on which the plant is grown outdoors. However, to generate optimum quantities of THC, the plant needs fertile soil and long hours of daylight, preferably in a sunny and warm climate. This means essentially that for the outdoor production of cannabis, growth occurs optimally anywhere within 35 degrees of the Equator. Typical growing regions include Mexico; northern India; and many areas of Africa, Afghanistan, California, and some southern US states.

Cannabis and Hemp

The cannabis plant is nowadays thought of mainly in the context of the psychoactive drug THC, but it is a versatile species that has had a very important place in human agriculture for thousands of years (for review, see Robinson, 1996). An acre of hemp produces more cellulose than an acre of trees, and the tough fiber produced from the outer layers of the stem has had many important uses. Hemp fiber made the ropes that lifted up the tough hemp-derived canvas cloth (the word derives from the Dutch pronunciation of cannabis) used to make the sails of the ancient Phoenician, Greek, and Roman navies. Archeological evidence shows that hemp fiber production was going on in northeastern Asia in Neolithic times, around 600 BC, and hemp production spread throughout the world, including the United States, where it was introduced by the first settlers. Although the importance of hemp declined with years, there were still 42,000 acres cultivated in the United States in 1917. Other major commercial centers of production were in Europe and Russia. Ships' sails, ropes, clothing, towels, and paper were all derived from hemp fiber and the woody cellulose-rich interior "hurds" of the hemp stem. Until the 1880s, almost all of the world's paper was made from hemp, and even today many bank notes are still printed on cannabis paper because of its toughness and durability. Most of our great artwork is painted on canvas, and the first Levi's jeans were made from canvas cloth.

Hemp seed has also been an important food crop, and from it can be derived an oil that has many uses as a lubricant, paint ingredient, ink solvent, and cooking oil. The seeds are now used mainly for animal feed and as birdseed.

Colorado and a number of other states have legalized the agricultural production of hemp, and this is now an increasingly popular "cash crop" in the United States. The cultivation of hemp as a fiber crop also continues in Europe. With the sanction of the European Union, farms in Hampshire in southern England, a traditional center for hemp farming, continue to grow the crop.

Chapter 2

The Pharmacology of
Delta-9-Tetrahydrocannabinol (THC)

Discovery of THC

As cannabis came into widespread use in Western medicine in the 19th century, it soon became apparent that the effects of plant-derived preparations were erratic. The amounts of active material that the pharmaceutical preparations contained were variable from batch to batch according to the origin of the material, the cultivation conditions, and the plant variety. Cannabis imported from India often deteriorated en route or in storage. Because the chemical identity of the active ingredients was not known and there was no method of measuring them, there was no possibility of quality control. This was one of the reasons why cannabis preparations eventually fell out of favor with physicians on both sides of the Atlantic. These inadequacies, however, also motivated an active research effort to identify the active principles present in the plant preparations in the hope that the pure compound or compounds might provide more reliable medicines. The 19th century was a great era for plant chemistry. Many complex drug molecules, known as alkaloids, were isolated and identified from plants. Several of these were powerful poisons—for example, atropine from deadly nightshade (*Atropa belladonna*); strychnine from the bark of the tree *Nux vomica*; and muscarine from the magic mushroom, *Amanita muscaria*. Others were valuable medicines still in use today—for example, morphine isolated from the opium poppy, *Papaver somniferum*;

the antimalarial drug quinine from the bark of the South American cinchona tree; and cocaine from the leaves of the coca plant. Victorian chemists were attracted to the new challenge offered by isolating the active ingredient from cannabis and attacked the problem with vigor, but initially without any notable success. Unlike the previously discovered alkaloids, which were all water-soluble organic bases that could form crystalline solids when combined with acids, the active principle of the cannabis plant proved to be almost completely insoluble in water. The active compound is in fact a viscous resin with no acidic or basic properties, so it cannot be crystallized. Because most of the previous successes of natural product chemistry had depended on the ability of chemists to extract an active drug substance from the plant with acids or alkalis and to obtain it in a pure crystalline form, it was not surprising that all of the early efforts to find the cannabis alkaloid in this way were doomed to failure. Only those who recognized that the active principle could not be extracted into aqueous solutions but required an organic solvent (usually alcohol) were able to make any real progress. T. and H. Smith, brothers who founded a pharmaceutical business in the mid-19th century in Edinburgh based on medicinal plant extracts, described in 1846 how they extracted Indian ganja repeatedly with warm water and sodium carbonate alkali to remove the water-soluble plant materials and then extracted the remaining dried ganja residue with absolute alcohol. The alcoholic extract was treated successively with alkaline milk of lime and with sulfuric acid and then evaporated to leave a small amount of viscous resin (6% or 7% of weight of the starting material) to which they gave the name "cannabin." It was clear from the nature of the procedures used that the resin was neither acid nor base but neutral. The purified resin proved to be highly active when tested in the then traditional manner on themselves: "Two thirds of a grain (44 mg) of this resin acts upon ourselves as a powerful narcotic, and one grain produces complete intoxication" (Smith and Smith, 1846).

The British chemists Wood, Spivey, and Easterfield, working in Cambridge at the end of the 19th century, made another important advance (see review by Todd, 1946). They used Indian charas (cannabis resin) as their starting material and extracted this with a mixture of alcohol and petroleum ether. From this, by using the then new technique of fractional distillation, they isolated a variety of different materials, including a red oil or resin of high boiling point (265°C) that was toxic in animals and that they suspected to be the active ingredient; they named it "cannabinol" (see Figure 1.2). A sample of the purified material was passed to the Professor of Medicine in Cambridge for further pharmacological investigation. The report published in *Lancet* in 1897 by his research assistant Dr. C. R. Marshall illustrates the heroic nature of pharmacological research in that era. Marshall described his experience on taking a sample of the material as follows:

> On the afternoon of Feb 19th last, whilst engaged in putting up an apparatus for the distillation of zinc ethyl, I took from 0.1 to 0.15 gramme of the pure substance from the end of a glass rod. It was about 2.30 p.m. The substance very gradually dissolved in my mouth; it possessed a peculiar pungent, aromatic, and slightly bitter taste, and seemed after some time to produce a slight anaesthesia of the mucous membranes covering the tongue and fauces. I forgot all about it and went on with my work. Soon after the zinc ethyl had commenced to distil—about 3.15—I suddenly felt a peculiar dryness in the mouth, apparently due to an increased viscidity of the saliva. This was quickly followed by paraesthesia and weakness in the legs, and this in turn by diminution in mental power and a tendency to wander aimlessly about the room. I now became unable to fix my attention on anything and I had the most irresistible desire to laugh. Everything seemed so ridiculously funny; even circumstances of a serious nature were productive of mirth. When told that a connection was broken and that air[1] was getting into the apparatus and an explosion feared I sat upon the stool and

1. Zinc ethyl burns on contact with air and consequently must be distilled in an atmosphere of carbon dioxide.

laughed incessantly for several minutes. Even now I remember how my cheeks ached. Shortly afterwards I managed to collect myself sufficiently to aid in the experiment, but I soon lapsed again into my former state. This alternating sobriety and risibility occurred again and again, but the lucid intervals gradually grew shorter and I soon fell under the full influence of the drug. I was now in a condition of acute intoxication, my speech was slurring, and my gait ataxic. I was free from all sense of care and worry and consequently felt extremely happy. When reclining in a chair I was happy beyond description, and afterwards I was told that I constantly exclaimed, "This is lovely!" But I do not remember having any hallucinations: The happiness seemed rather to result from an absence of all external irritation. Fits of laughter still occurred; the muscles of my face being sometimes drawn to an almost painful degree. The most peculiar effect was a complete loss of time relation: Time seemed to have no existence: I appeared to be living in a present without a future or a past. I was constantly taking out my watch thinking hours must have passed and only a few minutes had elapsed. This, I believe, was due to a complete loss of memory for recent events. Thus, if I walked out of the room I should return immediately, having completely forgotten that I had been there before. If I closed my eyes I forgot my surroundings and on one occasion I asked a friend standing near how he was several times within a minute. Between times I had merely closed my eyes and forgotten his existence. (Marshall, 1897, p. 238)

Marshall's colleagues became increasingly worried about him and eventually sent for medical help, but by the time the doctor arrived at approximately 5 p.m., Marshall had begun to recover, and by 6 p.m. he was on his way home after a cup of coffee and suffered no ill effects afterwards. Despite his experience, Marshall volunteered to take another dose of the resin three weeks later, but this time a much smaller one (50 mg). This produced essentially the same symptoms but in a milder form. It is clear that the red oil isolated by

Wood and colleagues was highly enriched in the active component or components of cannabis, and Marshall's description accurately describes the typical intoxication seen after high doses of the drug.

Although Wood and colleagues in Cambridge had come close to purifying the active ingredient in cannabis, their further work led them down a blind alley. From the "red oil" they were able to isolate a crystalline material after the preparation was acetylated [which produced acetyl derivatives of any compound with a free hydroxyl (–OH) group]. After purification of this crystalline derivative and removal of the acetyl groups by hydrolysis, they succeeded in isolating a compound that they called "**cannabinol**," and they showed that it could apparently be extracted from various other cannabis products, including several of the cannabis-containing medicines then available. The earlier red oil fraction was now renamed "crude cannabinol." Unfortunately, however, cannabinol was not the active ingredient but, rather, a chemical degradation product either formed during the chemical purification procedures or present as a normal degradation product in samples of cannabis material that had been stored for too long. The findings made with the original "red oil" material must have been due to the presence of THC in such samples. It was believed, erroneously, for decades after this that cannabinol was indeed considered to be the active principle of cannabis, although other laboratories were unable to repeat the findings of Wood et al.

Thirty years later, a brilliant young British chemist, Cahn, revisited the problem of cannabinol (see review by Todd, 1946). He was able to isolate the pure substance as described by Wood and colleagues, and using the improved chemical techniques available in the 1920s, he carried out a meticulous series of experiments that largely established the chemical structure of cannabinol (see Figure 1.2). Although this was not the true active principle, the new structure allowed chemists to synthesize a range of related compounds, and Cahn's work provided a great impetus to further chemistry research in this field.

At the University of Illinois in the 1940s, Roger Adams was also working on the problem (Adams, 1942). He used an alcoholic extract from which he produced "red oil" by distillation. From this he was able to purify a crystalline benzoic acid derivative of a compound that he named "cannabidiol" (because it contained two hydroxyl groups) and to work out its chemical structure (see Figure 1.2). This was a real advance because this compound— unlike the cannabinol worked on by Wood and colleagues—really is one of the naturally occurring materials in the cannabis plant. Unfortunately, however, it is not the active ingredient, and the narcotic activity that was reported by volunteers who took samples of Adams's cannabidiol must have been due to contamination with THC. Nevertheless, Adams and his group were able to synthesize various chemical derivatives of cannabidiol, including hydrogenated derivatives, the **tetrahydrocannabinols**, and some of these did possess potent psychoactive properties (measured both in human volunteers and increasingly by observing the behavioral responses of rodents, dogs, and other laboratory animals). In his 1942 "Harvey Lecture," Adams wrote,

> The typical marijuana activity manifested by the isomeric tetrahydrocannabinols constitutes ponderable evidence that the activity of the plant itself, and of extracts prepared therefrom, is due in large part to one or other of these compounds.

At the same time, across the Atlantic, despite the privations of war, research on cannabinoids continued in the Chemistry Department at Cambridge under the leadership of an outstanding organic chemist, Alexander Todd, later to become Lord Todd. He and his colleagues re-isolated cannabinol, and capitalizing on the newly discovered structure of cannabidiol published by the Adams group, they were able to complete the identification of the chemical structure of this compound started by Cahn (Todd, 1946). Both the Adams group and the Todd group went on to undertake the first chemical synthesis of cannabinol, and as part of this synthesis the Cambridge team actually made delta-9-tetrahydrocannabinol as an intermediate. They commented on the high degree of biological activity that this compound possessed (assessed by observing the characteristic behavioral reactions of dogs and rabbits rather than human subjects). The Todd group repeatedly tried to prove that this compound or something like it existed naturally in cannabis extracts. By repeated fractionation, they were able to prepare a highly active and almost colorless glassy resin that closely resembled synthetic tetrahydrocannabinol in its physical and chemical properties. The techniques available then, however, were not powerful enough to determine whether this was a single chemical substance or a complex mixture of closely related compounds. In a review article published in 1946, Todd wrote, "It would appear to be established that the activity of hemp resin, in rabbits and dogs at least, is to be attributed in the main to tetrahydrocannabinols."

THC was also isolated from a red oil fraction by the American chemist Wollner in 1942 although not as a single pure compound but, rather, as a mixture containing tetrahydrocannabinols. It was assumed for many years after the advances of the 1940s that the psychoactive properties of cannabis were due to an ill-defined mixture of such compounds. It was to be another 20 years before the brilliant chemical detective work of two Israeli scientists, Mechoulam and Gaoni, finally solved the problem and showed that in fact there is only one major active component, THC (Figure 2.1; Mechoulam, 1970). Mechoulam described his introduction to this field as follows:

> When we started our then very small programme on hashish some 5–6 years ago, our interest in this fascinating field was kindled by the contrast of rich folklore and popular belief with paucity of scientific knowledge. Israel is situated in a part of the world where, for many, hashish is a way of life. Though neither a producer nor a large consumer, Israel is a crossroads for smugglers, mostly Arab

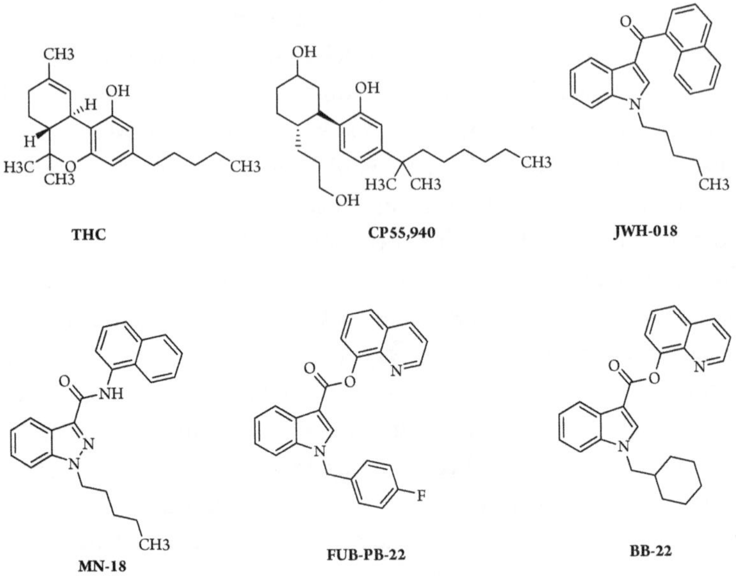

FIGURE 2.1 Some CB-1 selective antagonists.

Source: From Pertwee (2006).

Bedouin, who get Lebanese hashish from Jordan through the Negev and Sinai deserts to Egypt. Hence the police vaults are full of material waiting for a chemist.

Gaoni and Mechoulam had the advantage of new chemical separation and analytical techniques that had not been available to earlier investigators. In the Laboratory of Natural Products at the Hebrew University in Jerusalem, they had the latest methods for separating complex mixtures of chemicals by column chromatography. In this technique, the mixture is poured onto a column of adsorbent material and gradually washed through by solvents. Individual compounds move down the column at different rates according to how easily they dissolve in the solvent flowing through the column. In addition, the Israeli scientists were able to employ the powerful new techniques of mass spectrometry, infrared spectroscopy, and nuclear magnetic resonance to identify the chemicals that they had separated by chromatography. In this way, they were able to identify a large number of new cannabinoids in extracts of Lebanese hashish—we now know that as many as 60 different naturally occurring cannabinoids exist. Although this complexity might appear daunting, it turned out that most of the naturally occurring cannabinoids were present in relatively small amounts or that they lacked biological activity. In fact, Gaoni and Mechoulam (1964) reported that virtually all of the pharmacological activity in hashish extracts could be attributed to a single compound **delta-9-tetrahydrocannabinol (THC)**.[2]

Among other chemicals in the hashish extracts, Gaoni and Mechoulam (1964) identified **cannabidiol** as a major component (see Figure 2.1). They found a variety of other naturally occurring cannabinoids, but delta-9-THC was the most important. Cannabidiol is present in significant quantities but lacks psychoactive properties, although it may have other pharmacological effects (discussed later). Cannabis grown in tropical areas of the world (Africa, Southeast Asia, Brazil, Colombia, and Mexico) usually has much more THC than cannabidiol, with ratios of THC:cannabidiol of 10:1 or higher. Plants grown outdoors in more northern latitudes (Europe, Canada, and the northern United States), however, usually have a much higher content of cannabidiol, often exceeding the THC content by 2:1 (Clarke, 1981, p. 159). Cannabis also contains variable amounts of carboxylic acid derivatives of delta-9-THC, and this is potentially important. Although themselves inactive, the carboxylic acid derivatives readily lose their carboxylate group as carbon dioxide on heating, which gives rise to additional active

---

2. In some publications, including those from the Israeli group, this is referred to as delta-1-tetrahydrocannabinol, but this is because there are two different conventions for numbering the chemical ring systems of which the substances are composed; the delta-9 terminology is the most commonly used.

THC. This occurs, for example, when the plant material is heated during smoking or heated in the cooking processes used to form various cannabis-containing foods and drinks. This can in some instances more than double the active THC content of the original starting plant material. On the other hand, when cannabis resin or other preparations are stored, pharmacological activity is gradually lost and THC degrades by oxidation to cannabinol and other inactive materials.

The isolation and elucidation of the structure of delta-9-THC led to a burst of chemical synthetic activity throughout the world as different laboratories competed to be the first to complete the synthesis of this important new natural product. The American chemists Taylor, Lenard, and Shvo were probably the first in 1967, but they were quickly followed by Gaoni and Mechoulam and by several other laboratories (Mechoulam and Hanus, 2000). The Israeli group had shown that the naturally occurring THC occurred only as the *l*-isomer, although early synthetic preparations contained a mixture of both the *l*- and *d*- optical isomers (mirror images) of the compound. So the next stage was for several laboratories to devise chemical synthetic methods that yielded only the naturally occurring *l*-isomer of delta-9-THC, which is biologically far more active than the mirror-image *d*-isomer.

In retrospect, although the isolation of THC from cannabis proved technically difficult because of the nature of the compound as a neutral, water-insoluble, viscous resin, the outcome was not very different from that seen with other pharmacologically active substances derived from plants. In each case, a single active compound has been identified that accounts for virtually all of the biological activity in the crude plant extracts. This active compound often exists in the plant as one member of a complex mixture of related chemicals, most of which are either minor components or lack biological activity. This is true, for example, for nicotine from the tobacco leaf, cocaine from the coca leaf, and morphine from the opium poppy.

Other Phytocannabinoids

The cannabis plant contains more than 100 cannabinoids (ElSohly and Mahomood, 2010; ElSohly and Gul, 2015). Most of these are present in minor amounts in the plant. Turner et al. (2017) provided a detailed review of the pharmacology of phytocannabinoids. THC is both the most abundant and the most psychoactive compound. However, another important component is the non-psychoactive substance cannabidiol, present in approximately equal amounts to THC (see Figure 1.2). Cannabidiol is of pharmaceutical interest as a possible treatment for epilepsy and because it reduces the psychotomimetic actions of THC (Izzo et al., 2009; Niesink and van Laar, 2013; Iseger and Bossong, 2015). Cannabidiol has only weak affinity for the cannabinoid receptors, so its modulatory effects must be exerted elsewhere (Turner et al., 2017). It has been suggested that cannabidiol and one of the other phytocannabinoids, delta-9-tetrahydrocannabivarin, act as negative modulators of the endocannabinoid system (McPartland et al., 2015).

Synthetic Cannabinoids

For a review of synthetic cannabinoids, see Pertwee (2005, 2010).

The synthesis of THC was followed by a much larger chemistry effort, aimed at the synthesis and discovery of more potent analogs of THC, or compounds that separated the desirable medical properties of THC from its psychoactive effects. Many hundreds of new THC derivatives were made during the 1950s and 1960s in both academic and pharmaceutical company laboratories. There were far too many to be tested on human volunteers, so most were assessed in simple animal behavior tests that had been found to predict cannabis-like activity in humans (see Chapter 3). This research effort was disappointing because it proved impossible to separate the desirable properties of THC (anti-nausea and pain relieving) from the intoxicating effects. Nevertheless, the chemical research provided detailed insight into the "structure activity" of the THC molecule—that is, which parts of the molecule are critical for psychoactivity and which parts are less important and can thus be chemically modified without losing biological activity. Several derivatives proved to be even more active than THC, working in animals and human volunteers at doses up to 100 times lower than required for THC (for review, see Compton et al., 1993).

At the Pfizer company in the United States, for example, chemists were among the first to discover the first potent synthetic THC analog nantradol, which entered pilot scale clinical trials and was found to have analgesic (pain-relieving) properties that were not blocked by the drug naloxone—an antagonist that blocks analgesics of the morphine type which act on opiate receptors. Nantradol as synthesized originally was a mixture of four chemical isomers from which the active one, levonantradol, was later isolated. These compounds had an important advantage over THC in being more water soluble and thus easier to formulate and to deliver as a potential medicine. Further chemical work at Pfizer led to the discovery of a new chemical series of simplified THC analogs that possessed only two of the three rings of THC. Among these bicyclic compounds was the potent analog CP-55,940 (see Figure 2.1), which has been widely used as a valuable research compound. The Pfizer compound levonantradol was tested in several clinical trials during the early 1980s. It proved to be as potent as morphine as an analgesic, and it was effective in blocking nausea and vomiting in patients undergoing cancer chemotherapy, but the psychoactive side effects proved to be unacceptable and the company decided to abandon further research on this project (Dr. Ken Coe, personal communication).

Work in Raphael Mechoulam's laboratory in Israel was particularly productive in generating new analogs of THC (e.g., HU-210, which has particularly high affinity for both CB-1 and CB-2 receptors) (Mechoulam and Hanu, 2000). Research at the pharmaceutical company Eli Lilly in the United States led to the synthesis of

nabilone, the only synthetic THC analog that has been developed and approved as a medicine, sold under the trade name Cesamet (see Chapter 5).

In an unexpected development, research scientists at the Sterling Drug Company in the United States unwittingly discovered another chemical class of molecules which did not immediately resemble THC, but nevertheless proved to act through the same biological mechanisms. A research programme aimed at discovering novel aspirin-like anti-inflammatory/pain-relieving compounds generated an unusual lead compound called **pravadoline**. This had a remarkable profile in animal tests—it was highly effective in a broad range of pain tests—including ones in which aspirin-like molecules generally do not work. In addition, it failed to cause any gastric irritation, one of the major drawbacks in the aspirin class of drugs. Nor was pravadoline very effective in the key biochemical test for aspirin-like activity—the ability to inhibit the synthesis of the inflammatory chemicals, the prostaglandins. It seemed to the scientists involved that they had discovered a promising new mechanism for pain relief—and one that might have important advantages. Pravadoline went into clinical development, and meanwhile many other analogs were synthesized. From these emerged the compound WIN-55,212-2, an even more potent pain-relieving compound with improved absorption properties. However, when the specific receptor for cannabis was discovered in the 1980s (discussed later), it became clear that pravadoline and WIN-55,212-2 acted like THC on this receptor (Kuster et al., 1993), and were thus pharmacologically cannabinoids rather than aspirin-like anti-inflammatory drugs. Their pain-relieving properties were not due to a new mechanism but, rather, to the same mechanism as that of cannabis. Pravadoline had by that time been tested in human volunteers and found to possess good effectiveness against moderate to severe pain, such as postoperative dental pain. But it also caused dizziness and "light-headedness" as an obvious limiting side effect. The development of pravadoline was dropped because of kidney toxicity, and the company then decided to abandon the whole program—partly for budget reasons and partly to avoid being associated with the image of a "cannabis-like" drug (Dr. Susan Ward, personal communication).

While interest in the development of cannabis-based medicines waned in the pharmaceutical industry, academic research in this field remained active. From the synthesis and testing of many hundreds of chemical analogs, a consistent body of evidence was built up that defined the chemical structure–activity rules that determine whether a molecule will be active as a cannabimetic (for reviews, see Makriyannis and Rapaka, 1990; Compton et al., 1993; Mechoulam and Hanu, 2000; Thakur et al., 2005; Wiley et al., 2014).

This extensive academic research on the structure activity of cannabinoids, together with the discovery of the cannabinoid CB-1 receptor as their prime target, laid the basis for the unexpected hijacking of this field to generate chemicals that mimicked the actions of cannabis that could be sold to recreational users as "legal highs." Sold under the generic names "Spice" or "K2," these products were promoted originally as herbal materials, a form of "potpourri," "room incense," or as "liquid incense." Smoking mixtures are made to look like cannabis in brands that aspire to be "cannabis-like," but high potency brands tend to look artificial and are often dyed a bright color. At first, people believed that Spice was simply a mixture of harmless herbs that had a similar effect to cannabis. It was sold legally throughout the world, especially via the internet, attractively packaged in small colorful sachets. In fact, some astute research in Germany in 2008 revealed that the innocuous herbal mixture was laced with small amounts of synthetic cannabinoids; the previously known agonist JWH-018 (see Figure 2.1) was one of the first to be detected.

Many of the synthetic cannabinoids monitored by the EMCDDA [European Monitoring Centre for Drugs and Drug Addiction] through the EU Early Warning System have code names that relate to their discovery. In some cases they are derived from the initials of the name of the scientists that first synthesised them: e.g. "JWH" compounds after John W. Huffman and "AM" compounds after Alexandros Makriyannis. In other cases code names may originate from the institution or company where they were first synthesised, the "HU" series of synthetic cannabinoids being from the Hebrew University in Jerusalem, or "CP" from Carl Pfizer. In some cases names have probably been chosen by those making "legal high" products to help market the products. Striking examples of this are "AKB-48" and "2NE1," alternative names used for APINACA and APICA. "AKB-48" is the name of a popular Japanese girl band and "2NE1" is the name of a girl band from South Korea. Finally, the synthetic cannabinoid XLR-11, appears to have been named after the first liquid fuel rocket developed in the USA for use in aircraft, perhaps alluding to the vendor's intention for those who consume the substance. Many synthetic cannabinoids are now given code names that are derived from their long chemical names, such as APICA from N-(1-adamantyl)-1-pentyl-1H-indole-3-carboxamide, and APINACA from N-(1-adamantyl)-1-pentyl-1H-indazole-3-carboxamide. (EMCDDA, 2016a)

Spice became popular among users in Europe and the United States in the early years of the century and soon led to government attempts to ban it. Unfortunately, the extensive published research and the relative promiscuity of the CB-1 receptor meant that cannabinoids of several chemical classes could be used. In the United Kingdom, the Advisory Council on the Misuse of Drugs (an independent group that advises government) recommended the control of a wide range of known synthetic cannabinoids in 2009, but the Council soon found that underground chemists were able to find ways around this control, and the control of a "second generation" of compounds was recommended in 2012, followed by the recommendation for the control of a "third generation" of compounds in 2014, updated in 2016 (Advisory Council on the Misuse of Drugs, 2009, 2012, 2014, 2016). Several hundred synthetic cannabinoids have been made, and reports of tests show that most have an affinity for the CB-1 receptor similar to that of THC and some are substantially more potent than THC (De Luca et al. 2016; Hess et al., 2016). The "third-generation" compounds BB-22 and FU-BB-22 (see Figure 2.1) are approximately 30 times more potent than JWH-018, with binding affinities for CB-1 receptor as low as 0.1 nM. In addition, whereas THC is a "partial agonist" at the CB-1 receptor, the synthetic cannabinoids behave as full agonists (ElSohly et al., 2014; Hess et al., 2016). With each new generation of synthetic cannabinoids, there was a tendency for increased potency; some of the synthetic substances are far more potent than cultivated herbal cannabis ("skunk").

The synthetic cannabinoids continue to be popular on both sides of the Atlantic. Abuse of these drugs reached epidemic proportions in many British prisons. The liquid forms can be used in vaporizers, and the synthetic drugs are difficult to detect without complex analytical equipment, allowing users to evade the drug testing regimes now routinely used to detect cannabis use. The manufacture of novel illegal psychoactive compounds by using the results of published academic research is not unique to the cannabinoids. Similar examples can be seen in the use of scientific publications to generate novel opiates, Ritalin-like drugs, ketamine analogs, and benzodiazepines (Schifano et al., 2015). Users are offered compounds that have not been through the extensive safety testing to which new human medicines are normally subject; their widespread use is a form of "Russian roulette," hoping that serious adverse effects do not occur.

Cannabinoid Antagonists

An important development has been the discovery of molecules that bind to the cannabis receptor in brain but instead of mimicking THC, they block its actions. Like the synthetic cannabinoids, these come from various different chemical classes, and three examples are shown in Figure 2.2.

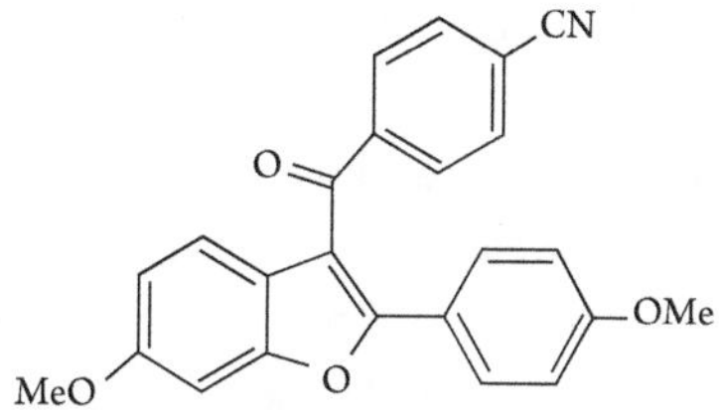

LY320135

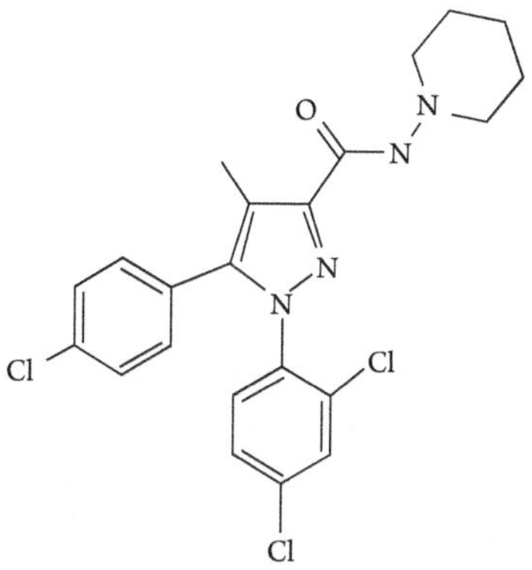

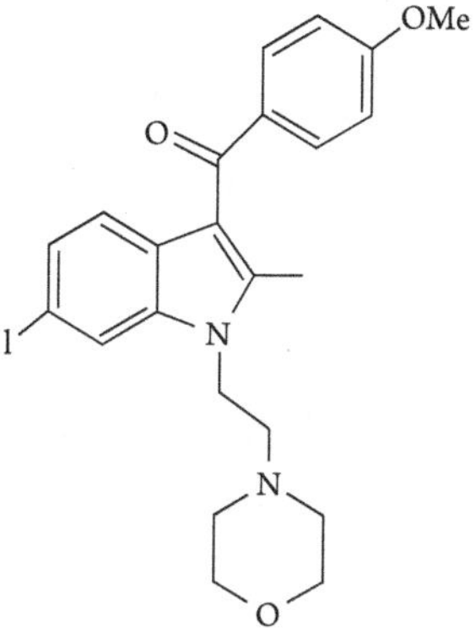

SR141716A   AM-630

FIGURE 2.2 Synthetic drugs that act as antagonists at the CB-1 cannabinoid receptor. SR141716A rimonabant.

The first cannabinoid antagonist to be described was the compound SR141716A, now called rimonabant, from the French pharmaceutical company Sanofi-Aventis (see Figure 2.2). Rimonabant has been used extensively both as a valuable research tool and as a potential medicine. It entered large-scale clinical trials for the treatment of metabolic disease and obesity (see Chapter 5). The potential of rimonabant for treating obesity prompted others to develop CB-1 receptor antagonists including the rimonabant analogs AM 251, AM 281, and AM 630; the Lilly compound LY320135 (see Figure 2.2);

and the very potent Chinese compound MJ15 (K_i = 27.2 pM) (Pertwee, 2006). In addition to blocking the actions of CB-1 agonists, these compounds also behave as "inverse agonists." In the absence of added agonists, they elicit responses opposite to those induced by agonists. This is thought to be due to the suppression of the "constitutive" activity of the CB-1 receptor (Pertwee, 2005, 2006). Compounds were also developed that act as selective antagonists at CB-2 receptors, notably SR144528, JTE907, and AM630. These compounds bind with much higher affinity to CB-2 than to CB-1 receptors, they are potent CB-2 antagonists, and they behave as inverse agonists that by themselves produce cannabimimetic effects at CB-2 receptors (Pertwee, 2006).

CB-2 Selective Agonists

Most of the synthetic cannabinoids described previously are active at both CB-1 and CB-2 receptors. There has been interest in CB-2 selective compounds for therapeutic use because such compounds are devoid of the psychoactive intoxicant effects that limit the usefulness of CB-1 agonists (Manera et al., 2016). Pertwee (2006) listed several such compounds (Table 2.1); notable were the potent

TABLE 2.1 CB-1 Binding Profiles: [H^3]CP-55,940 Assay of Rat Brain Membranes

Drug	K_p Concentration for Half Occupancy of Receptor Binding Sites—Nanomolar (10^{-9} M)
(−)CP-55,940	0.068
(+)CP-55,940	3.4
THC	1.6
11-Hydroxy-THC	1.6
Cannabinol	13.0
Cannabidiol	>500.0

Source: Devane et al. (1988).

and selective CB-2 agonist HU-308 and the antagonists/inverse agonists SR144528 and AM30, both at least 100 times more potent against CB-2 than against CB-1 receptors.

How Does THC Get to the Brain?

For a review on this subject, see Huestis (2007).

Smoking

Smoking is an especially effective way of delivering psychoactive drugs to the brain (Huestis et al., 1992). When marijuana is smoked, some of the THC in the burning plant material distils into a vapor (the boiling point of THC is approximately 200°C), and as the vapor cools, the compound condenses again into fine droplets, forming a smoke that is inhaled. As the drug dissolves readily in fats, it passes readily through the membranes lining the lungs, which offer a large surface area for absorption. The drug enters blood, which passes directly from the lungs to the heart, from where it is pumped in the arteries around the body. THC has no difficulty penetrating into the brain, and within seconds of inhaling the first puff of marijuana smoke, active drug is present on the cannabis receptors in the brain. Peak blood levels are reached at approximately the time that smoking is finished (Figure 2.3).

Smoking is one of the most efficient ways of rapidly experiencing the effects of cannabis on the brain (see Chapter 4). The favorite of many people in the West is the marijuana "**joint**." This consists of herbal cannabis (from which stems and seeds have first carefully been removed), rolled inside a rice paper cylinder either by hand or using a rolling machine. A typical joint contains approximately 0.5 g of leaf. Europeans favor "**spliffs**," with added tobacco—which assists the otherwise often erratic burning of the marijuana. Many different slang words describe herbal marijuana, including "Aunt

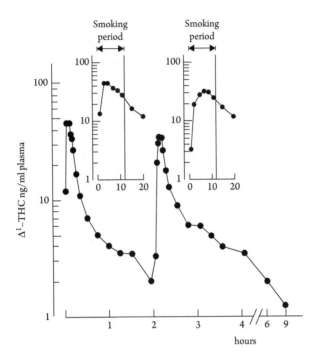

FIGURE 2.3 Average blood levels of THC in human volunteers who smoked two identical marijuana cigarettes, each containing approximately 9 mg of THC, 2 hours apart. Insets show the rapid absorption of the drug during the period of smoking.

Source: From S. Agurell, M. Halldin, J. E. Lindgren, A. Ohlsson, M. Widman, H. Gillespie, and L. Hollister (1986). Pharmacokinetics and metabolism of delta 1-tetrahydrocannabinol and other cannabinoids with emphasis on man. *Pharmacol Rev.* 38(1):21–43.

Mary," "dope," "grass," "Mary Jane," "reefer," and "weed." When a joint or spliff has been smoked down to the point that it is difficult to hold, it is called a "roach," and this still contains appreciable amounts of THC that gradually distils down the length of the joint as it is smoked. The roach may be held in the split end of a match or with a variety of "roach pins" or tweezers with which one may hold the roach without burning oneself. Another smoking device is the "**blunt**," which consists of a cigar from which tobacco has been removed and replaced with marijuana, or the blunt is rolled with tobacco paper and filled with marijuana. Blunts contain more marijuana than joints or spliffs, and they are best suited to smoking in social gatherings. In the social groups in which marijuana is commonly smoked, as with the port served in Oxford and Cambridge Colleges after dinner, etiquette demands that the joint or blunt is passed around the group in a circular fashion. As with the port, hoarding of the joint by any one person is regarded as a serious breach of protocol. Experienced marijuana smokers often develop the technique of inhaling a considerable quantity of air along with the smoke—this dilutes the smoke, making it less irritating to the airways and allowing deeper inhalation.

Marijuana smoker can regulate almost on a puff-by-puff basis the dose of THC delivered to the brain to achieve the desired psychological effect and to avoid overdose and minimize the undesired effects. Puff and inhalation volumes tend to be higher at the beginning and lowest at the end of smoking a cigarette (more drug is delivered in the last part of the joint because some THC condenses onto this). When experienced smokers were tested with joints containing different amounts of THC (from 1% to 4%), without knowing which was which, they adjusted their smoking behavior to reach about the same level of THC absorption and subjective high. When smoking the less potent cigarettes, puff volumes were larger and puff duration higher than with the more potent cigarettes. In addition, when smoking the latter, more air was inhaled, thereby diluting the marijuana smoke (Herning et al., 1986). These results, however, were not confirmed by Chait (1986). The amount of THC absorbed by smoking varies over quite a large range. Of the total amount of THC in a marijuana cigarette, on average approximately 20% will be absorbed, with the rest being lost by combustion, sidestream smoke, and incomplete absorption in the lung; however, the actual figure ranges from less than 10% to more than 30% even among experienced smokers.

Many marijuana smokers hold their breath for periods of 10–15 seconds after inhaling, in the belief that this maximizes the subjective response to the drug. However, studies in which subjective responses and THC levels in blood were measured with different breath-hold intervals have failed to show that breath-holding makes any real difference to the absorption of the drug—this idea thus seems to fall in the realms of folklore rather than reality.

Marijuana can also be smoked using a variety of pipes. A simple pipe resembling those used for tobacco can be used, but marijuana pipes are usually made of such heat-resistant materials as stone, glass, ivory, or metal. This is necessary because marijuana does not tend to stay alight in a pipe, so it constantly has to be relit. A common variety of pipe is the water pipe or "bong"; more complex water pipes are known as "hookahs." Pipes come in many different forms, but all use the same principle. Smoke from the pipe is sucked through a layer of water, which cools it and is thought to remove much of the tar and other irritant materials present in herbal marijuana smoke, although the effectiveness of this practice is debatable. Bongs tend to be complex and heavy devices and thus not easily portable.

Pipe smoking is the traditional method for smoking marijuana ("ganja") in India and hashish in the Arab world. Khwaja A. Hasan gives the following description of ganja smoking in contemporary India:

> Ganja is smoked in a funnel-shaped clay pipe called chilam. Almost anybody except the untouchables (sweeper caste) can join the group and enjoy a few puffs. The base part of the bowl portion of the funnel-shaped pipe is first covered with a small charred clay filter. Then the mixture of ganja and tobacco is placed on this filter. A small ring, the size of the bowl, of rope fibre called baand is first burnt separately and then quickly placed on top of the smoking material. The pipe is now ready for smoking. Usually four or five people gather around a pipe. . . . Ritual purity of the pipe is always preserved for the clay pipe is never touched by the lips of the

smoker. The tubular part of the chilam at its bottom is held in the right hand and the left hand also supports it. The passage between the index finger and the thumb of the right hand is used in taking puffs from the pipe. . . . While they sit in a squatting position on a chabootra (raised platform) in front of one person's house, or gather in an open space while the host prepares the chilam they talk about social problems, weather, crops, prices, marriage negotiations and so forth. Such gatherings may take place at any time during the day except early morning. After a smoke they again go back to work. Thus such smoking parties are like "coffee breaks" in the American culture. (Hasan as cited in Rubin, 1975)

It is clear why smoking is the preferred route of delivery of cannabis for many people. As with other psychoactive drugs, the rapidity by which smoking can deliver active drug to the brain and the accuracy with which the smoker can adjust the dose delivered are powerful pluses. The rapid delivery of drug to the receptor sites in the brain seems to be an important feature in determining the subjective experience of the high. This is true not only for cannabis but also for other psychoactive drugs that are smoked, including nicotine, crack cocaine, methamphetamine, and, increasingly, heroin ("chasing the dragon"). For the narcotic drugs, smoking is the only method that approaches the instant delivery of drug achieved by intravenous injection—and it does not carry the risks of infection with hepatitis or HIV associated with intravenous use. Fortunately, the extreme insolubility of cannabis makes use of the intravenous route very difficult.

Vaping

Advances in vaporization technology have offered smokers an alternative method with fewer health concerns. The effects associated with smoking are widely debated (see Chapter 6), but health professionals are in agreement that smoke-free methods pose less risk of bronchial irritation and are medically preferred. A wide variety of "vaporizers" are available. They all aim to heat the marijuana sample to 180°–200°C, which causes the THC and other cannabinoids to vaporize so that they can then be inhaled. Heating is ether directly through a "hot plate" or indirectly by heating air to the correct temperature. The temperature should be kept to less than 365°C to avoid volatilizing tar or other unwanted components. Users may use buds or resin, but increasingly vaporizers are used to heat and volatilize a small sample of one of the many concentrates now available. A popular procedure is to use a drop of "hash oil," a concentrate containing a high THC content. Discrete vaporizers in the form of "pens" no larger than a fountain pen, fuelled with drops of hash oil, allow the consumer to go undetected in public places. Inhaling THC vapor avoids the possible adverse effects of smoke on the lungs, but it is quite expensive. When using concentrates, the user may find it more difficult to regulate the dose. Although smoking may avoid the dangers inherent in injectable drug use, it is still associated with the potential damage that may be caused to the lungs. Vaping avoids such hazards because herbal cannabis is not burnt, so users do not inhale the potentially hazardous products of combustion. Vaping is increasingly popular, and in Canada, a majority of medical cannabis users prefer this route (Shiplo et al., 2016).

Dabbing

Whereas the use of smokeless vaporizers can be discrete, the growing popularity of "dabbing" is far from this. The users employ a variety of distinct devices to heat a small sample of solid concentrate or oil (a "dab") and then to inhale the vapors. The devices are often complex, expensive pieces of glassware, with the heating element being a nail that is often heated by the user with a blowtorch before adding the sample to the red-hot surface. Possibly with the legalization of cannabis in several states, this public display of cannabis use is a kind of defiant gesture to society at large. The use of concentrates also makes it difficult for the user to regulate dose.

Oral Absorption

Taking THC by mouth is unreliable as a method of delivering a consistent dose of the drug. THC is absorbed reasonably well from the gut, but the process is slow and unpredictable, and most of the absorbed drug is rapidly degraded by metabolism in the liver before it reaches the general circulation. The peak blood levels of THC occur anywhere between 1 and 4 hours after ingestion, and the overall delivery of active THC to the bloodstream averages less than 10%, with a large range between individuals. The "high" is correspondingly also delayed by comparison with smoking (Figure 2.4).

Even for the same person, the amount of drug absorbed after oral ingestion will vary according to whether the person has eaten a meal recently and the amount of fat in the food. A further complication of the oral route is that one of the metabolites formed in the liver is 11-hydroxy-THC (Figure 2.5). This is a psychoactive metabolite with potency approximately the same as that of THC. The amount of 11-OH-THC formed after smoking is relatively small (plasma levels are less than one-third of those for THC), but when cannabis is taken by the oral route—in which all the blood from the intestine must first pass through the liver—the amount of 11-OH-THC in plasma is approximately equal to that of THC, and it probably contributes at least as importantly as THC to the overall effect of the drug.

The heating of cannabis leads to the formation of additional THC from the chemical breakdown of pharmacologically inactive carboxylic acid THC derivatives present in the plant preparations (see Chapter 1). Despite this unpredictability, THC is soluble in fats and in alcohol, so it can be extracted and added to various foodstuffs and drinks. Originally, edibles were limited to homemade brownies that were unappetizing and contained a variable dose of THC. Nowadays, one can find medicated cookies, popcorn, crackers, nut mixes, lollipops, ice cream, gummy bears, chocolate bars, chews,

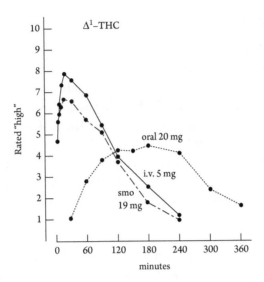

FIGURE 2.4 Time course of the subjective "high" after administering THC by different routes. Smoking gives as rapid an effect as an intravenous injection, whereas taking the drug by mouth produces a delayed and prolonged high. The subjective experience somewhat outlasts the presence of THC in blood (see Figure 2.5) because THC persists longer in the brain.

Source: From S. Agurell, M. Halldin, J. E. Lindgren, A. Ohlsson, M. Widman, H. Gillespie, and L. Hollister (1986). Pharmacokinetics and metabolism of delta 1-tetrahydrocannabinol and other cannabinoids with emphasis on man. *Pharmacol Rev.* 38(1):21–43.

and many other kinds of food. Culinary science has evolved enough that most products are appetizing; you can hardly tell that they contain cannabis. One danger is that children may find such products attractive, so users need to keep them secure. The oral route avoids the irritant effects of inhaled smoke that many people find objectionable. The delayed and prolonged absorption are attractive to patients taking marijuana chronically for medical use. A common method is to heat the plant leaf in butter, margarine, or cooking oil and then strain out the solid plant materials and use the oil or butter for cooking.

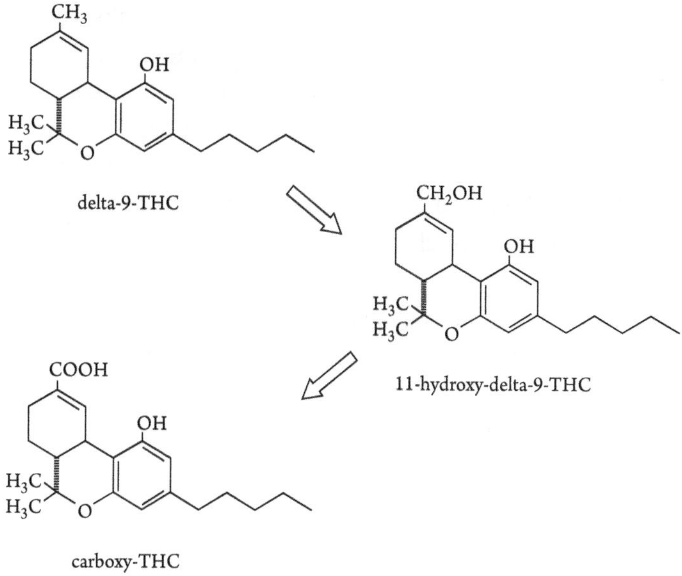

FIGURE 2.5 Principal route of metabolism of THC.

Other Routes of Administration

THC can be extracted with alcohol by heating and straining, yielding a variety of tinctures that can be diluted with lemonade or other flavored drinks. In the former US and British medical use of cannabis, the formulations used were alcoholic extracts of the plant, sometimes diluted further with alcohol to yield "tincture of cannabis." This was diluted with water and administered by mouth. Cannabis tinctures are available for application by spray under the tongue (sublingual tinctures). The only herbal cannabis product currently approved for medical use, nabiximols (Sativex), is administered in this way (see Chapter 5). This route is used for a variety of traditional medicines (e.g., nitroglycerine) and appears to be effective. Medical users of cannabis also employ transdermal patches and the local application of cannabis-containing salves, ointments, lotions, and sprays for arthritis, chapped skin, eczema, minor burns, muscle soreness, sunburns, swellings, joint pain, and tendonitis. Another way of delivering the drug is in the form of a rectal suppository. Absorption from the rectum bypasses the liver and avoids the problem of liver metabolism, which limits the oral availability of THC, and it seems that this route can deliver approximately twice as much active drug to the bloodstream as the oral route, although there is still considerable variability in drug absorption from one individual to another. Nevertheless, the slow and prolonged absorption is attractive to some patients seeking chronic treatment with cannabis for medical conditions (see Chapter 5), although administration is awkward and sometimes embarrassing.

Elimination of THC from the Body

For a review on how THC is eliminated from the body, see Huestis (2007).

After smoking, blood levels rise very rapidly and then decline to approximately 10% of the peak values within the first hour (see Figure 2.3). The maximum subjective high is also attained rapidly and persists for approximately 1 or 2 hours, although some milder psychological effects last several hours. After oral ingestion, the peak for plasma THC and the subjective high is delayed and may occur anywhere from 1 to 4 hours after ingestion, with mild psychological effects persisting for up to 6 hours or more (see Figure 2.4). Although in each case unchanged THC disappears quite rapidly from the circulation, elimination of the drug from the body is in fact quite complex and takes several days. This is largely because the fat-soluble THC and some of its fat-soluble metabolites rapidly leave the blood and enter the fat tissues of the body. They are then slowly released back into other body compartments, including the brain. As the drug and its metabolites are gradually excreted in the urine (approximately one-third) and in the feces (approximately two-thirds), this gives an overall elimination "half-time" of 3–5 days. Cannabinoids are metabolized in the liver. A major metabolite is 11-hydroxy-THC, which is possibly more potent than THC itself and may be responsible for some of the effects of cannabis (see Figure 2.5).

More than 20 other metabolites are known, some of which are psychoactive and all of which have long half-lives of several days, and some drug metabolites may persist for several weeks after a single drug exposure (Agurell et al., 1986). Urine or blood tests for one of the major metabolites, the inactive 11-nor-carboxy-THC (see Figure 2.5), use a very sensitive immunoassay and can give positive results for days or even weeks after a single drug exposure. Such measurements form the basis of cannabis urine tests to determine whether or not an individual has consumed cannabis. Even after a single dose of cannabis, the user may test positive several days later, and regular cannabis users may remain positive for up to a month after taking the last dose. What is more relevant for roadside traffic accident or workplace drug testing is an indication of recent use or intoxication, and this is better provided by the measurement of THC in samples of saliva, which accurately reflect cannabis use within the past few hours (Erowid, n.d.). With the increasing medical and recreational use of cannabis, a simple test that could be administered at the roadside to determine whether a driver is intoxicated with cannabis is increasingly attractive. A number of technical advances permit the sensitive measurement of THC in saliva (NarcoCheck, 2015). In Britain since March 2015, police have been issued with a roadside test kit that can detect cannabis (THC), cocaine, and heroin based on immunoassays. Although measuring THC is far more relevant than previous roadside tests that measured the carboxy-THC metabolite, there is little agreement on where the upper limit of THC concentration in saliva should be set. Some US states have set a limit of 5 ng THC/ml saliva, but others suggest 25 ng THC/ml saliva, and this is still contentious. Drivers may differ in their susceptibility to THC, and there is no firm agreement about the adverse effects of cannabis on driving. For people caught with positive cannabis tests, often applied randomly in the workplace or because they were involved in road traffic accidents or were admitted to hospital emergency rooms, the consequences can be serious.

The unusually long persistence of THC in the body has given cause from some concern, but it is not unique to THC; it is seen also with a number of other fat-soluble drugs, including some of the commonly used psychoactive agents such as diazepam (Valium). The presence of small amounts of THC in fat tissues has no observable effects because the levels are very low. There is no evidence that THC residues persist in the brain, and the slow leakage of THC from fat tissues into blood does not give rise to drug levels that are high enough to cause any psychological effects. Smoking a second marijuana cigarette a couple of hours after the first generates virtually the same plasma levels of THC as previously (see Figure 2.3). Nevertheless, the drug will tend to accumulate in the body if it is used regularly. Although this is not likely to be a problem for occasional or light users, there have been few studies of chronic high-dose cannabis users to determine whether the increasing amounts of drug accumulating in fat tissues could have harmful consequences. Is it possible, for example, that such residual stores of drug could sometimes give rise to the "flashback" experience that some cannabis users report—the sudden recurrence of a subjective high not associated with drug taking?

How Does THC Work?

Discovery of Cannabinoid Receptors

Pharmacologists used to think that the psychoactive effects of cannabis were somehow related to the ability of the drug to dissolve in the fat-rich membranes of nerve cells and disrupt their function. But the amount of drug that is needed to cause intoxication is exceedingly small. An average marijuana joint of 0.4–0.6 g of herbal "buds" contains 40–100 mg of THC (1 mg = 0.001 g, or approximately 0.0003 of an ounce). (Cannabis is usually purchased as one-eighth of an ounce (3.5 g) or as 50-g samples—the latter enough for approximately 100–200 joints.) Less than half of the total THC content is absorbed by the smoker, so an average total body dose may be between 20 and 50 mg of THC. The amount of drug ending up in the brain, which accounts for only approximately 2% of total body weight, can be predicted to be not more than 1,000 μg (1 μg = 0.000001 g). Although these are exceedingly small amounts, they are comparable to the naturally occurring amounts of other chemical compounds used in various forms of chemical signaling in the brain. The brain works partly as an electrical machine, transmitting pulses of electrical activity along nerve fibers connecting one nerve cell to another, but the actual transmission of the signal from cell to cell involves the release of pulses of chemical signal molecules known as "neurotransmitters." These chemicals are specifically recognized by receptors, which are specialized proteins located in the cell membranes of target cells. The neurotransmitter chemicals are released in minute quantities: For example, the total amount of one typical neurotransmitter, noradrenaline, in the human brain is not more than 100–200 μg—a quantity comparable to the intoxicating dose of THC. This suggests that THC most likely acts by targeting one or other of the specific chemical signaling systems in the brain rather than by some less specific effect on nerve cell membranes, and indeed this is what the scientific evidence indicates.

An important breakthrough in understanding the target on which THC acts in the brain was the discovery by Allyn Howlett and colleagues at St. Louis University in 1986 of a biochemical model system in which THC and the new synthetic cannabinoid drugs WIN-55,12-2 and CP-55,940 were active (for review, see

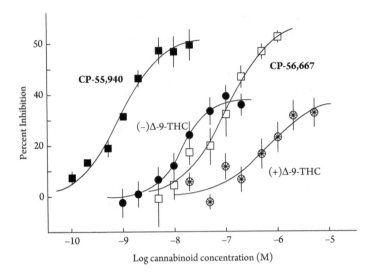

FIGURE 2.6 Inhibition of cyclic AMP formation in tissue culture cells that possess the CB-1 cannabinoid receptor. The synthetic cannabinoid CP-55,940 is more potent than (–)delta-9-THC and produces a larger maximum inhibition. The response shows selectivity for the (–) isomers of the compounds versus the (+) isomers (CP-56,667 is the (+) isomer of CP-55,940).

Source: From Matsuda et al. (1990).

Howlett, 2005; see also Abood and Martin, 1992; Mechoulam et al., 2014). The cannabinoids were found to inhibit the activity of an enzyme in rat brain, adenylate cyclase, which synthesizes a molecule known as cyclic AMP (Figure 2.6). The significance of this finding was that the synthesis of cyclic AMP is known to be controlled by a number of different cell surface receptors that recognize neurotransmitter substances. Some receptors when activated stimulate cyclic AMP formation, whereas others inhibit it. Cyclic AMP is known as a "second messenger" molecule because some chemical messengers form it inside cells in response to activation of a receptor at the cell surface. Cyclic AMP acts as an important control molecule inside the cell, regulating many different aspects of cell metabolism and function. Thus, Howlett's discovery suggested that she had found an indirect way to study drug actions on the "cannabis receptor" in brain.

A few years later in 1988, Howlett's group went one step further and found a more direct way to study drug actions at the "cannabis receptor." A popular method for studying drug actions at cell surface receptors is to measure the selective binding of a substance known to act specifically on such a receptor to the receptor sites in fragments of brain cell membranes incubated in a test tube. In order to measure the very small amounts of drug bound to the receptors—which are only present in small numbers—the drug molecule is usually tagged by incorporating a small amount of radioactivity into the molecule. The radioactive drug can then be measured very sensitively by radioactive detection equipment. Sol Snyder and colleagues at Johns Hopkins University in Baltimore pioneered the application of this method to the study of drug receptors in brain during the 1970s. In a now famous experiment, Snyder and his student Candice Pert used a radioactively labeled derivative of morphine to show that rat brain possessed specific "opiate receptors" that selectively bound this and all other pharmacologically active opiate drugs (Pert and Snyder, 1973)). This experimental approach was subsequently used to devise binding assays for all of the known neurotransmitter receptors in brain and peripheral tissues. Such assays offer a simple method for determining whether any compound interacts with a given receptor, and they provide a precise estimate of its potency by measuring the concentration needed to displace the radiolabeled tracer.

Snyder's group and several others had tried to determine whether a binding assay could be devised for the "cannabis receptor" by incubating rat brain membranes with radioactively labeled THC in a test tube. This failed, however, because the THC dissolves very readily in the lipid-rich cell membranes—and this nonspecific binding to the membranes completely obscured the tiny amount of radiolabeled THC bound specifically to the receptors. Howlett collaborated with research scientists at the Pfizer pharmaceutical company to solve this problem. They achieved success by using not THC but, rather, the synthetic compound CP-55,940 discovered at the company laboratory as a very potent synthetic cannabinoid (see Figure 2.2). This had the advantage of being even more potent than THC—and thus binding even more tightly to the cannabis receptor. In addition, because CP-55,940 is more water soluble than THC, there was much less nonspecific binding of the radiolabeled drug to the rat brain membrane preparations. The binding assay that resulted seemed faithfully to reflect the known pharmacology of THC and various synthetic cannabinoids (Table 2.1; Devane et al., 1988). Thus, THC and the psychoactive metabolite 11-hydroxy-THC were able to displace radiolabeled CP-55,940 at very low concentrations— approximately 1 nm (equivalent to <1 μg in a liter of fluid—and compatible with the amounts of THC thought to be present in brain after intoxicant doses). Cannabidiol and other cannabinoids were inactive, and the d-isomer of CP-55,940—known to be much less potent in animal behavior models—was approximately 50 times less potent in displacing the radiolabel than the more active mirror-image l-isomer. The binding assay was quickly adopted and was used, for example, to confirm that the Sterling–Winthrop compound WIN-55,212-2 acted specifically at the cannabis receptor; indeed, radio-actively labeled WIN-55,212-2 could be used as an alternative label in binding studies to identify the cannabis receptor. Another facet of cannabis pharmacology was emphasized by the discovery of these biochemical models, namely that the cannabis receptor seemed to be a wholly novel discovery—not related in any obvious way to any of the previously known receptors for neurotransmitters in brain. None of the neurotransmitters themselves, or the other chemical modulators in brain, the neuropeptides, interacted to any extent in the cannabis binding assay (Table 2.1).

Research on the cannabinoid receptor has shown that THC acts as "partial agonist" at the CB-1 receptor—that is, it is not able to elicit the full activation of the receptor seen, for example, with the synthetic compounds CP-55,940 and WIN-55,212-2. This can be shown by the fact that THC does not cause the same maximum inhibition of adenylate cyclase as the synthetic compounds (see Figure 2.6).

An alternative functional assay measures the ability of various agonists to stimulate the binding of a metabolically stabilized analog of GTP (GTP-γ-S) to the activated receptor. In this assay, THC is also only partly effective (25–30%) by comparison with the synthetic cannabinoids. This assay also reveals some level of "constitutive" activity in the CB-1 receptor—reflected by binding of GTP-γ-S in the absence of any added cannabinoid. This is inhibited by the antagonist rimonabant, suggesting that in addition to its ability to antagonize the actions of cannabinoid agonists, this compound also acts as an "inverse agonist" at the CB-1 receptor (Pertwee, 2005). The activity of other CB-1 and CB-2 antagonists as inverse agonists has been noted previously. It is not clear whether such baseline receptor activity is great enough to be of any physiological significance, but if so, it might explain some of the pharmacological effects that have been observed in animals when treated with the antagonists alone.

The cannabis receptor in brain belongs to a family of related receptor proteins, and in 1990 a group working at the US National Institutes of Health isolated the gene encoding it (Matsuda et al., 1990). This provided independent confirmation of the unique nature of the cannabis receptor. A few years later, a second gene was discovered that encoded a similar but distinct subtype of cannabis receptor, now known as the CB-2 receptor, to distinguish it from the CB-1 receptor in brain (for review, see Felder and Glass, 1998). CB-2 receptors also bind radioactively tagged CP-55,940 and recognize most of the cannabinoids that act at CB-1 sites. The CB-2 receptor, however, is clearly different and is found mainly in peripheral tissues, particularly on white blood cells—the various components of the immune system of the body—although it is also present at low levels in some brain regions, notably the brainstem (Van Sickle et al., 2005). It may be that actions of THC on peripheral CB-2 sites may account for some of the effects of cannabis on the immune system. Research on CB-2 receptors has been helped by the availability of selective antagonists at these receptors (e.g., SR144528) and the development of CB-2 selective synthetic agonists (e.g., AM630 and AM1241) (Pertwee, 2006) (Table 2.2). Studies of the functional roles of CB-1 and CB-2 receptors have also been greatly helped by the development of genetically modified strains of mice, in which the expression of one or other receptor has been "knocked out" (Valverde et al., 2005; see Chapter 3, this volume). The CB-1 receptors also contain "allosteric" sites, distinct from the agonist binding sites, which modulate receptor function. A number of small molecule allosteric modulators, both positive and negative, are known, and these could offer an alternative approach to modifying CB-1 receptor function (Pertwee, 2006). Studies of the purified human CB-1 receptor using X-ray crystallography have revealed its detailed molecular structure and have provided insights into the way in which THC and synthetic cannabinoids bind to the receptor (Hua et al., 2016; Shao et al., 2016).

It is possible that further cannabinoid receptors remain to be discovered. By searching DNA sequence databases for the characteristic seven-membrane spanning feature of receptors of the cannabinoid type, more than 140 novel "receptors" have been described (GPCRs). One of these, GRP55, is activated by THC and endocannabinoids and may be a third cannabinoid receptor (Ryberg

TABLE 2.2 Some Selective CB-2 Receptor Agonists

CB-2 Selective Agonist	CB-1, K_i (nM)	CB-2, K_i (nM)
AM-1240	280	3.4
JWH-133	677	0.4
GW-405833	477	0.9
JWH-015	383	3.8
HU-308	>10,000	2.7

Source: Pinder (2007).

et al., 2007). However, GRP55 does not closely resemble the other cannabinoid receptors, although its distribution in central nervous system and peripheral tissues is similar (Schicho and Storr, 2012). It seems more likely that GRP55 and the family of related receptors are activated normally by phospholipids related to lysophosphatidic acid (Oka et al., 2010; Marichal-Cancino et al., 2017). The therapeutic potential of GPCRs and GRP35 and CRP55 and their relation to chemokines and phospholipids are reviewed by Shore and Reggio (2015).

It is notable that the other abundant naturally occurring cannabinoid, cannabidiol, interacts only weakly with either the CB-1 or the CB-2 receptor. Nevertheless, cannabidiol does possess some pharmacological activities that do not appear to be related directly to actions on CB-1 or CB-2 receptors (Cascio and Pertwee, 2015; Turner et al., 2017). Cannabidiol is active in some animal models of epilepsy (see Chapter 5). Some animal and human psychopharmacology experiments suggest that cannabidiol may have therapeutic effects in a variety of disorders, including diabetes, gastrointestinal disturbances, cancer, oxidative stress, inflammation, and cardiovascular disease (Sullan et al., 2017). Cannabidiol is legally available in cannabis plants bred for high levels of cannabidiol and also as an oil suitable for vaping, and it is used in self-medication for epilepsy, although the mechanism of action remains unknown. It has been suggested that cannabidiol acts as a negative allosteric modulator of CB-1 receptor function (McPartland et al., 2015).

Cannabinoids can also activate the "transient receptor potential vanilloid 1" (TRPV1) receptor (Tabrizi et al., 2017). TRPV1 is an ion channel expressed on sensory neurons. Its activation, triggered by a number of agonists including cannabinoids, increases sensitivity to both chemical and thermal stimuli. TRPV1 functions as a sensor of noxious stimuli, and it has been associated with chronic pain conditions.

Research on the cannabinoid receptor has shown that THC acts as "partial agonist" at the CB-1 receptor—that is, it is not able to elicit the full activation of the receptor seen, for example, with the synthetic compounds CP-55,940 and WIN-55,212-2. This can be shown by the fact that THC does not cause the same maximum inhibition of adenylate cyclase as the synthetic compounds (see Figure 2.6). An alternative functional assay measures the ability of various agonists to stimulate the binding of GTP-γ-S to the activated receptor. In this assay, THC is also only partially effective (25–30%) by comparison with the synthetic cannabinoids. This assay also reveals some level of "constitutive" activity in the CB-1 receptor— reflected by binding of GTP-γ-S in the absence of any added cannabinoid. This is inhibited by the antagonist rimonabant, suggesting that in addition to its ability to antagonize the actions of cannabinoid agonists, this compound also acts as an "inverse agonist" at the CB-1 receptor (i.e., it can block the resting level of activity in the receptor that occurs in the absence of cannabinoid agonists). It is not clear whether such baseline receptor activity is great enough to be of any physiological significance, but if so, it might explain some of the pharmacological effects that have been observed in animals when treated with the antagonist alone (Pertwee, 2005).

Neuroanatomical Distribution of CB-1 and CB-2 Receptors in Brain and Peripheral Tissues

The distribution of cannabinoid receptors was first mapped in rat brain in autoradiographic studies, using the radioligand [H^3]CP-55,940, which binds with high affinity to CB-1 sites. This method involves incubating thin sections of brain tissue with the radiolabeled drug and subsequently using a photographic emulsion sensitive to radiation to detect where the radiotracer was selectively bound (Herkenham et al., 1991) (Figure 2.7).

The validity of using this radioligand was confirmed by autoradiographic studies in CB-1 receptor knockout mice (genetically engineered so that they do not have any CB-1 receptors in their brains), in which no detectable [H^3]CP-55,940 binding sites were

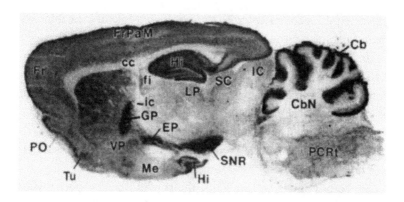

FIGURE 2.7 Distribution of cannabinoid CB-1 receptor in rat brain revealed by an autoradiograph of the binding of radioactively labeled CP-55,940 to a brain section. The brain regions labeled are as follows: Cb, cerebellum; CbN, deep cerebellar nucleus; cc, corpus callosum; EP, entopeduncular nucleus; fl, fimbria hippocampus; Fr, frontal cortex; FrPaM, frontoparietal cortex motor area; GP, globus pallidus; Hi, hippocampus; IC, inferior colliculus; LP, lateral posterior thalamus; Me, medial amygdaloid nucleus; PO, primary olfactory cortex; PCRt, parvocellular reticular nucleus; SNR, substantia nigra pars reticulata; Tu, olfactory tubercle; and VP, ventroposterior thalamus.

Source: Photograph kindly supplied by Dr. Miles Herkenham, National Institute of Mental Health, USA.

observed (Zimmer et al., 1999). Subsequently, antibodies that target particular regions of the CB-1 receptor protein were used for immunohistochemical mapping studies (Egertová and Elphick, 2000). Immunohistochemistry provides a superior degree of spatial resolution compared to autoradiography, and it allows very high-resolution mapping at the electron microscope level, but the overall pattern of distribution of CB-1 receptors revealed by the two approaches proved to be very similar (Mackie, 2005).

The mapping studies in rat brain showed that CB-1 receptors are mainly localized to axons and nerve terminals and are largely absent from the neuronal cell bodies or dendrites. The finding that cannabinoid receptors are predominantly presynaptic rather than postsynaptic is consistent with the postulated role of cannabinoids in modulating neurotransmitter release.

In the brain, the CB-1 receptors are abundant in the cerebellum, basal ganglia, hippocampus, and dorsal primary afferent spinal cord regions, which is why cannabinoids influence functions such as memory processing, pain regulation, and motor control. In the brainstem, the concentration of cannabinoid receptors is low, which may be related to why cannabis use is not associated with sudden death due to depressed respiration, for example.

In both animals and humans, the cerebral cortex, particularly frontal regions, contains high densities of CB-1 receptors. Mapping of CB-1 receptors in human brain has revealed a very similar anatomical distribution to that seen in laboratory animals (Mackie, 2005).

CB-1 receptors also exist in substantial densities in some peripheral tissues, notably liver and adipose tissue, in which they are thought to play an important role in metabolic control and the regulation of food intake (see Chapter 3) (Matias and Di Marzo, 2007).

CB-2 receptors are largely confined to white blood cells in lymph nodes and also other tissues of the immune system. They are present at much lower densities than CB-1 receptors in the brain, where they are mainly seen on glial cells, although some reports suggest a neuronal localization also (Szabo, 2015).

Chapter 3

Peripheral and Central Effects of THC

Inhibition of Neurotransmitter Release

For more detailed reviews of the inhibition of neurotransmitter release, see Iversen (2003) and Pertwee (2015a).

Although we have only a limited knowledge of how activation of the CB-1 receptor in brain leads to the many actions of delta-9-tetrahyrocannabinol (THC), some general features of cannabinoid control mechanisms are emerging. CB-1 receptors are often coupled to inhibition of cyclic AMP formation, but this is not always the case. In some nerve cells, activation of CB-1 receptors inhibits the function of calcium ion channels, particularly those of the N-subtype. This may help explain how cannabinoids inhibit the release of neurotransmitters, because these channels are essential for the release of these substances from nerve terminals. CB-1 receptors are not usually located on the cell body regions of nerve cells—where they might control the electrical firing of the cells—but are concentrated instead on the terminals of the nerve fibers, at sites where they make contacts (known as synapses) with other nerve cells. Here, the CB-1 receptors are well placed to modify the amounts of chemical neurotransmitter released from nerve terminals and thus to modulate the process of synaptic transmission by regulating the amounts of chemical messenger molecules released when the nerve terminal is activated. Experiments with nerve cells in tissue culture or with thin slices of brain tissue incubated in the test tube have shown that the addition of THC or other cannabinoids can inhibit the stimulation-evoked release of various neurotransmitters, including the inhibitory amino acid GABA and the amines noradrenaline and acetylcholine (Szabo and Schlicker, 2005). In the peripheral nervous system, CB-1 receptors are also found on the terminals of some of the nerves that innervate various smooth muscle tissues (Mackie, 2005). Roger Pertwee and colleagues in Aberdeen have made use of this in devising a variety of organ bath assays, in which THC and other cannabinoids inhibit the contractions of smooth muscle in the intestine, vas deferens, and urinary bladder evoked by electrical stimulation. Such bioassays have proved valuable in assessing the agonist/antagonist properties of novel cannabinoid drugs (Pertwee, 2005). Although cannabinoids generally inhibit neurotransmitter release, this does not mean that their overall effect is always to dampen down activity in neural circuits. For example, reducing the release of the powerful inhibitory chemical GABA would have the opposite effect, by reducing the level of inhibition. This may explain two important effects of cannabinoids that have been described in recent years: that administration of THC leads to a selective increase in the release of the neurotransmitter dopamine in a region of brain known of the nucleus accumbens and that this is accompanied by an activation and increased release of naturally occurring opioids (endorphins) in brain (see Chapter 4). Whereas inhibition of the release of synaptic neurotransmitters is the mechanism of action of externally administered THC and other cannabinoids, synaptic pharmacology is far more complex for the naturally occurring endocannabinoids, whose synaptic release is triggered by locally released neurotransmitters and whose actions may be "retrograde" to inhibit further neurotransmitter release.

Effects on the Heart and Blood Vessels

For a review of the effects of THC on the heart and blood vessels, see Sullan et al. (2017).

The cannabinoids exert profound effects on the vascular system. Initial results in animals suggested that the main effect of THC and anandamide was to cause a decrease in heart rate and a lowering of blood pressure. However, it became apparent that the fact that experimental animals were anesthetized profoundly affected the results; in non-anesthetized animals and in humans, the predominant effect is an increase in heart rate and a lowering in blood pressure. This is due to the action of THC on the smooth muscle in the arteries, causing a relaxation that leads to an increase in their diameter (vasodilatation). This can be observed in isolated blood vessels, indicating that it is not an indirect effect mediated via the central nervous system (CNS). The vasodilation in turn leads to a drop in blood pressure as the resistance to blood flow is decreased, and this automatically triggers an increase in heart rate in an attempt to compensate for the decrease in blood pressure. The vasodilatation caused by THC in human subjects is readily seen as a reddening of the eyes caused by the dilated blood vessels in the conjunctiva. The cardiac effects can be quite large, with increases in heart rate in humans that can be equivalent to as much as a 60% increase over the resting pulse rate. Although this presents little risk to young healthy people, it could be dangerous for patients who have a history of heart disease, particularly those who have suffered a heart attack or heart failure previously. Another feature commonly seen after a high dose of cannabis is "postural hypotension"—that is, people are less able to adjust their blood pressure adequately when rising from a seated or lying down position. This leads to a temporary drop in blood pressure, which in turn can cause dizziness or even fainting.

It was initially assumed that the effects of the cannabinoids on the heart and blood vessels were mediated indirectly through actions on receptors in the brain. It is now clear, however, that some of these effects are mediated locally through CB-1 receptors located in the blood vessels and heart. Isolated blood vessels relax when incubated with the endocannabinoid anandamide, and this effect and the vascular effects in the whole animal can be blocked by the CB-1 antagonist rimonabant. This antagonist also blocks the effects of THC on blood pressure and heart rate in animals and humans. Furthermore, the cardiovascular effects of THC are completely absent in CB-1 receptor knockout mice. Other physiological effects of cannabinoids may also be due to direct actions on CB-1 receptors on blood vessels. These include the ability to lower the pressure of fluid in the eyeball (intraocular pressure)—an effect that underlies the medical use of cannabis in the treatment of raised intraocular pressure in glaucoma (see Chapter 4).

Effects on Pain Sensitivity

For reviews of the effects of THC on pain sensitivity, see Costa and Comelli (2015) and Woodhams et al. (2015).

Pain relief by cannabinoids (analgesia) is one of the key features in the "Billy Martin tetrad" (discussed later); it represents one of the most important potential medical applications for these substances (see Chapter 5). THC and synthetic cannabinoids are effective in many animal models of both acute pain (mechanical pressure, chemical irritants, and noxious heat) and chronic pain (e.g., inflamed joint following injection of inflammatory stimulus or sensitized limb after partial nerve damage). In all these cases, the pain-relieving effects of cannabinoids are prevented by co-treatment with the antagonist rimonabant and are absent in CB-1 receptor knockout mice, indicating that the CB-1 receptor plays a key role. CB-1 receptors are present in high densities at various relay stations in the neural pathways that transmit pain information into the CNS, including primary sensory pain-sensitive neurons, spinal cord, brainstem, and other relay sites. CB-1 receptors on the spinal cord and on primary sensory nerves' peripheral tissues are also thought to play an important role. Animals in which the CB-1 receptor was genetically ablated selectively in primary sensory neurons showed enhanced sensitivity to noxious heat or mechanical stimuli (Agarwal et al., 2007). This distribution is similar to that of the opiate receptor and the endogenous morphine-like brain chemicals—the endorphins. However, the opiate and cannabinoid systems appear to be parallel but distinct (Fields and Meng, 1998). Treatment of animals with low doses of naloxone, a highly selective antagonist of opioid receptor, completely blocks the analgesic effects of morphine but generally has little or no effect in reducing the analgesic actions of THC or other cannabinoids. Conversely, rimonabant generally has little or no effect on morphine analgesia. Nevertheless, there are links between these two systems. In some experiments, it has been found that cannabinoids and opiates act synergistically in producing pain relief—that is, the combination is more effective than either drug alone in a manner that is more than simply additive. For example, in the mouse tail-flick response to radiant heat and in a rat model of arthritis (inflamed joint), doses of THC that by themselves were ineffective made the animals more sensitive to low doses of morphine (Smith et al., 1998). Such synergism could have potentially useful applications in the clinic (see Chapter 5). This is potentially a way of avoiding the unacceptable psychoactivity of THC for the treatment of pain because only low doses of THC are required. It has generally been assumed that the site of action of the cannabinoids in producing pain relief is in the CNS. In support of the concept of a central site of action, several studies have shown that cannabinoids can produce pain relief in animals when they are injected directly into spinal cord or brain. However, there is also evidence for a dual action of cannabinoids at both CNS and peripheral tissue levels. In a rat inflammatory pain model, in which the irritant substance carrageenan was injected into a paw, injection of very small amounts of cannabinoids into the inflamed paw inhibited the development of increased pain sensitivity normally seen in this model, perhaps by acting on CB-1 receptors on primary sensory nerve endings. The peripheral injection of compounds selective for CB-2 receptors also caused analgesia in this model, and a CB-2 selective antagonist blocked these effects. CB-2 selective agonists are also effective in a variety of pain models, and their effects are not blocked by rimonabant. It is possible that the effects of the CB-2 selective compounds are due in part to their anti-inflammatory actions in suppressing immune system responses to injury. These findings suggest an important role for peripheral sites in mediating the overall analgesic effects of cannabinoids and point to potential future applications for topically administered cannabinoids and/or CB-2 selective agonists in pain control (Walker and Hohmann, 2005; Kunos and Tam, 2011; Shang and Tang, 2016). Another interesting observation is that the CB-1 antagonist rimonabant, in addition to blocking the analgesic effects of cannabinoids, may sometimes when given by itself make animals more sensitive to painful stimuli—that is, the opposite of analgesia. The simplest interpretation of this finding is that there may be a constant release of endogenous cannabinoids in pain circuits and that these compounds thus play a physiological role in setting pain thresholds. Alternatively, some of the CB-1 receptors in the body may have some level of activation even when not stimulated by cannabinoids; rimonabant might then act as an "inverse agonist" to suppress this receptor activity.

Effects on Motility and Posture

For a review of the effects of THC on motility and posture, see Fernández-Ruiz (2009).

Cannabinoids cause a complex series of changes in animal motility and posture. At low doses, there is a mixture of depressant and stimulatory effects, and at higher doses there is predominantly CNS depression. In small laboratory animals, THC and other cannabinoids cause a dose-dependent reduction in their spontaneous running activity. This may be accompanied by sudden bursts of activity in response to sensory stimuli—reflecting a hypersensitivity of reflex activity. Adams and Martin (1996, p. 1590) described the syndrome in mice as follows:

> Δ^9-THC and other psychoactive cannabinoids in mice produce a "popcorn" effect. Groups of mice in an apparently sedate state will jump (hyperreflexia) in response to auditory or tactile stimuli. As animals fall into other animals, they resemble corn popping in a popcorn machine.

At higher doses, the animals become immobile and will remain unmoving for long periods, often in unnatural postures—a phenomenon known as "catalepsy." Similar phenomena are observed in large animals. One of the first reports of the pharmacology of cannabis was published in the *British Medical Journal* more than 100 years ago (Dixon, 1899). Dixon described the effects of extracts of Indian hemp in cats and dogs as follows:

> Animals after the administration of cannabis by the mouth show symptoms in from three quarters of an hour to an hour and a half. In the preliminary stage cats appear uneasy, they exhibit a liking for the dark, and occasionally utter high pitched cries. Dogs are less easily influenced and the preliminary condition here is one of excitement, the animal rushing wildly about and barking vigorously. This stage passes insidiously into the second, that of intoxication. . . . In cats the disposition is generally changed showing itself by the animals no longer demonstrating their antipathy to dogs as in the normal condition, but by rubbing up against them whilst constantly purring; similarly a dog which was inclined to be evil-tempered and savage in its normal condition, when under the influence of hemp became docile and affectionate. . . . When standing they hold their legs widely apart and show a peculiar to and fro swaying movement quite characteristic of the condition. The gait is exceedingly awkward, the animal rolling from side to side, lifting

its legs unnecessarily high in its attempts to walk, and occasionally falling. A loss of power later becomes apparent especially in the hind limbs, which seem incapable of being extended. Sudden and almost convulsive starts may occur as a result of cutaneous stimulation, or loud noises. The sensory symptoms are not so well defined, but there is a general indifference to position. Dogs placed on their feet will stay thus till forced to move by their ataxia, whilst if placed on their side they continue to lie without attempting a movement. . . . Animals generally become more and more listless and drowsy, losing the peculiar startlings so characteristic in the earlier stage, and eventually sleep three or four hours, after which they may be quite in a normal condition.

Monkeys respond similarly to THC, with an initial period of sluggishness followed by a period of almost complete immobility. The animals typically withdraw into the far corner of the observation cage and adopt a posture that has been called the "thinker position" because the monkeys have a tendency to support their head with one hand and have a typical blank gaze. Human marijuana users may also sometimes withdraw from contact with other members of the group and remain unmoving for considerable periods of time.

These effects of cannabinoids most likely reflect their actions on CB-1 receptors in an area of brain known as the basal ganglia, which is importantly involved in the control and initiation of voluntary movements, and a region at the back of the brain known as the cerebellum, which is involved in the fine-tuning of voluntary movements and the control of balance and posture (Bostan and Strick, 2010). CB-1 receptors are present in some abundance in both of these brain regions.

The Billy Martin Tetrad

As chemical efforts to synthesize novel THC analogs and other synthetic cannabinoids intensified after the discovery of THC, it became increasingly important to have available simple animal tests that might help predict which compounds retained THC-like CNS pharmacology—in particular, which might be psychoactive in humans. Although it is never possible to determine whether an animal is experiencing intoxication, certain simple tests do seem to have some predictive value. Professor Billy Martin, at the Medical College of Virginia, devised a series of four simple behavioral tests that have been widely used (Martin, 1985). He demonstrated that drugs that produced in mice a combination of reduced motility, lowered body temperature, analgesia, and immobility (catalepsy) were very likely to be psychoactive in humans. The four symptoms are readily measured experimentally and exhibit dose-dependent responses to cannabinoids. By testing a large number of compounds in the "Billy Martin tetrad," Martin and colleagues showed that there was a good correlation between the potencies of the various cannabinoids in these tests and their affinities for the CB-1 receptor, as measured in a radioligand binding assay in the test tube (Figure 3.1). Furthermore, the CB-1 receptor antagonist rimonabant completely blocks all four responses.

Laboratory Studies of Cannabinoids in Human Volunteers

For a review of laboratory studies of cannabinoids, see Sherif et al. (2016).

The sudden popularity of marijuana use among young people during the 1960s in America prompted an upsurge of scientific research on the drug's effects. A large and often confusing literature emerged, partly because the topic was politically charged from the outset and bias undoubtedly colored some of the investigations. Some researchers seem to have been intent on proving that marijuana was a harmful drug. Others tended to emphasize the benign aspects of the drug, and some focused on the ability of cannabinoids

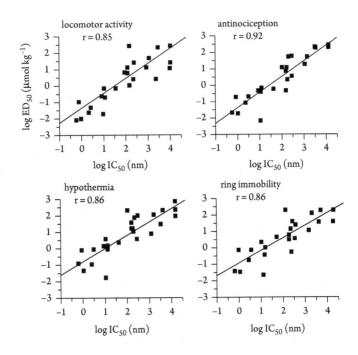

FIGURE 3.1 The Billy Martin tests. Correlations between in vivo and in vitro activities of more than 25 cannabinoid analogs to inhibit spontaneous activity (locomotor activity), reduce sensitivity to pain (tail-flick test; antinociception), reduce body temperature (hypothermia), and cause immobility (ring immobility) in mice are plotted against affinities of the same compounds for CB-1 receptors assessed in an in vitro binding assay using radioactively labeled CP-55,940.

Source: From Abood and Martin (1992).

to precipitate schizophrenia-like psychotic effects (Freedman et al., 2015; Sherif et al., 2016).

Studying a psychotropic drug under laboratory conditions is never easy. It is difficult, for example, to ensure that subjects receive a standard dose because of the inconsistent absorption of THC—even by regular users. Many of the early studies in the United States used illicit supplies of marijuana of dubious and inconsistent potency.

Standardized marijuana cigarettes eventually became available for academic research studies. They were produced for the National Institute on Drug Abuse using cannabis plants grown for the government agency by the University of Mississippi. When methods became available for measuring the THC content of the plant material, it was possible by judicious blending of marijuana of high and low THC content to produce marijuana cigarettes with a consistent THC content. By using plant material with low THC content or marijuana from which THC had been extracted by soaking in alcohol, "placebo" cigarettes with little or no THC could also be produced. It is nevertheless difficult to ensure that subjects are exposed to the same amounts of THC or other cannabinoids because of individual differences in absorption and smoking behavior. A more precise method is to deliver cannabinoids intravenously; this offers reliable pharmacokinetics and reduces inter-individual differences in absorption. Intravenous administration is difficult, however, because of the insolubility of THC and related substances in water or saline. Nevertheless, an intravenous solution can be achieved by dissolving the cannabinoid in dilute alcohol and a surface-active agent (THC was dissolved in a vehicle containing 0.4–1.0% Tween-80 and 0.4–1.0% ethanol in saline; Tanda et al., 2000), and researchers have used the intravenous route for numerous laboratory studies (reviewed by Englund et al., 2012).

Included in "laboratory studies" are those that use other routes of administration and ask specific questions and in which the results of subjects taking the drug are compared to those from a similar number of subjects using an inactive placebo (placebo-controlled studies). Ideally, to avoid possible bias, neither the researcher nor the subject should be aware of whether active drug or placebo is being administered (double-bind, placebo-controlled studies).

The question of how to select suitable human subjects for such studies is also difficult. The effects of marijuana in inexperienced or completely naive subjects taking it for the first time are very different from those seen in experienced regular drug users. In one of the very first controlled studies, carried out at Boston University, drug-naive subjects were compared with experienced users. As in many subsequent studies, the naive users showed larger drug-induced deficits in the various tasks designed to test cognitive and motor functions compared to drug-experienced subjects, who often show no deficits at all (Weil et al., 1968).

Effects on Learning and Memory

For reviews of the effects of TCH on learning and memory, see Solowij (1998) and Shrivastava et al. (2011).

There have been numerous studies of higher brain functions in human subjects given intoxicating doses of marijuana. The results have not always confirmed the subjective experiences of the subjects. Thus, although subjectively users report a heightened sensitivity to auditory and visual stimuli, laboratory tests fail to reveal any changes in their sensory thresholds. If anything, they become less sensitive to auditory stimuli. The feeling of heightened sensitivity must therefore involve higher perceptual processing centers in the brain rather than the sensory systems themselves. On the other hand, the perceived changes in the sense of time are readily confirmed by laboratory studies. Subjects are asked to indicate when a specified interval of time has passed or to estimate the duration of an interval of time generated by the investigator. In such tests, intoxicated subjects overestimate the amount of elapsed time (Matthew et al., 1998). Thus, marijuana makes people experience time as passing more quickly than it is in reality; in other words, marijuana increases the subjective time rate. One minute seems like several. This curious effect can also be seen in rats trained to respond for food reward using a fixed-interval schedule. When treated with THC or WIN-55,212-2, the animals shortened their response interval, whereas the antagonist rimonabant lengthened this interval (Han and Robinson, 2001).

Several studies have examined acute impairments in mental functioning and memory (Shrivastava et al., 2011). Such tests are administered under laboratory conditions, and nowadays the tests usually involve an interactive video screen and computerized test batteries (for review, see Robbins et al., 1994). In simple mental arithmetic tasks or repetitive visual or auditory tasks that require the subject to remain attentive and vigilant, marijuana seems to have little effect on performance, although if the task requires the subject to maintain concentration over prolonged periods of time (>30 minutes), performance falls off. By far the most consistent and clear-cut acute effect of marijuana is to disrupt short-term memory. "Short-term memory" is nowadays usually described as "working memory." It refers to the system in the brain that is responsible for the short-term maintenance of information needed for the performance of complex tasks that demand planning, comprehension, and reasoning ("executive function"). Working memory can be tested in many ways. In the expanded "digit span" test, subjects are asked to repeat increasingly longer strings of random numbers both in the order in which they are presented and backwards. In this test, marijuana has been reported to produce a dose-dependent impairment in most studies. Other tests involve the presentation of lists of words or other items, and subjects are asked to recall the list after a delay of varying interval. Again, people intoxicated with marijuana show impairments, and as in the digit span tests, they characteristically exhibit "intrusion errors"—that is, they tend to add items to the list that were not there originally. The drug-induced deficits in these tests become even more marked if subjects are exposed to distracting stimuli during the delay interval between presentation and recall. Marijuana makes it difficult for subjects to retain information on line in working memory in order to process it in any complex manner. This is consistent with the results of brain imaging studies that showed that increases in blood flow after THC administration to volunteers were greatest in the frontal cortex, and this is where there was the best correlation with subjective reports of intoxication

(Matthew and Wilson, 1993; Matthew et al., 1997). The frontal cortex contains the highest densities of CB-1 receptors of all cortical regions (Herkenham et al., 1991) and is known to be important in the control of executive brain function (i.e., coordinating information in short-term stores and using it to make decisions or to begin to lay down more stable memories). The hippocampus, another region enriched in cannabinoid receptors, interacts importantly with the cerebral cortex, particularly in visuospatial memory and in the processes by which working memory can be converted to longer term storage.

The complex effects of cannabis on cognitive function can be divided into the various periods following cannabis use (for review, see Crean et al., 2011):

1. Acute effects (0–7 hours): Short-term memory is clearly impaired along with measures of attention and concentration, accompanied by impairment on tasks measuring information processing, a fundamental building block for attention and concentration. The acute effects of cannabis can be interpreted as impairments in executive function (coordinating information in short-term stores and using it to make decisions or to begin to lay down more stable memories).

2. Residual effects of cannabis on executive function (7 hours–20 days after last use): Cannabis use may impair executive function for several weeks. In the period of "recent abstinence" (7 hours–20 days), users may experience impairments in attention, concentration, inhibition, and increased impulsivity and other aspects of executive function. No impairments in short-term memory have been found. Impairments in executive function are most severe in subjects who have been smoking large amounts of cannabis for long periods of time. The residual impairments are linked to the duration and quantity of cannabis use—one might consider these effects as a form of withdrawal after sudden abstinence.

3. Long-term effects of cannabis on executive functions (3 weeks or longer after last use): Does cannabis use lead to cognitive impairments that persist after drug use has stopped? This question has attracted a great deal of research attention. The scientific literature, however, is confused by differences in the definition of "long-term" use and inconsistent findings. It is possible that some of the impairments observed in the period of sudden abstinence from the drug are effects of "withdrawal" rather than genuine long-term effects of drug use. Nevertheless, in the few studies that avoid this complication, subtle impairments in executive function remain. Although basic intentional and short-term memory deficits are restored, there are enduring deficits in decision-making, concept formation, and planning. As previously noted, the effects are most marked in subjects with a previous history of chronic, heavy cannabis use (Crean et al., 2011).

Van Amsterdam et al. (1996) note the many methodological difficulties inherent in studies of the long-term consequences of marijuana use. Among the confounding factors in human studies are that comparisons have to be made between groups of drug users and non-users; it is usually impossible to compare the baseline performance of these groups prior to cannabis use to determine if they are properly matched. A cohort of more than 3,000 people in the United States was studied over a period of 25 years using three tests designed to assess verbal memory, processing speed, and executive function. Current use of cannabis was associated with defects in all three aspects of cognitive function. After excluding current users and adjusting for potential confounders, cumulative lifetime exposure to cannabis showed a progressive impairment in verbal memory, equivalent to remembering one word less from a list of 15 words for every 5 years of previous cannabis use (Auer et al., 2016). A review of the effects of cannabis on cognitive function, covering publications between 2004 and 2015, confirmed that verbal learning and memory, in addition to attention, were most impaired after acute or chronic exposure to cannabis. After prolonged abstinence, impaired verbal memory and attention and impairment of some aspects of executive function may persist (Broyd et al., 2016). On the other hand, a review of studies of cannabis users with at least 1 month of abstinence failed to find any deficit in performance on neuropsychological tests compared to that of non-users, after correction for confounding factors (Schreiner and Dunn, 2012).

Subjective Reports of the Marijuana High

A number of approaches can be used to study the effects of drugs on the brain. We can ask people taking the drug to report their own subjective experiences—and there is a large and colorful literature of this type on marijuana. Green et al. (2003) reviewed studies of self-reported cannabis experiences. They concluded that euphoria and relaxation were the primary positive effects of cannabis, although there was considerable variation in the results. Scientists prefer to use objective methods, and many experiments have been performed with human volunteers to determine what physiological and psychological alterations in brain function are induced by the drug. Studying the effects of the drug on animal behavior can also help us understand how the drug affects the human brain. Understanding how the drug acts in the brain and which brain regions contain the highest densities of drug receptors may also provide useful clues. We can be reasonably confident that the psychic effects of cannabis are due to activation of the CB-1 receptor in brain. Huestis et al. (2001) carried out a controlled study on 63 healthy cannabis users, who received either rimonabant or placebo and smoked either a THC-containing or placebo marijuana cigarette. The CB-1 antagonist blocked all of the acute psychological effects of the active cigarettes.

Millions of people take marijuana because of its unique psychotropic effects. It is difficult to make a precise scientific description of the state of intoxication caused by marijuana because this is clearly an intensely subjective experience not easily put into words, and the experience will vary enormously depending on many variables. Some of these are easily identified.

First, the *dose of the drug* is clearly important. It will determine whether the user merely becomes "high" (i.e., pleasantly intoxicated) or escalates to the next level of intoxication and becomes "stoned"— a state that may be associated with hallucinations and end with immobility and sleep. High doses of cannabis carry the risk of unpleasant experiences (panic attacks or even psychosis). Experienced users become adept at judging the dose of drug needed to achieve the desired level of intoxication, although this is much more difficult for naive users. The dose is also much easier to control when the drug is smoked; it is more difficult to control when taken by mouth.

Second, the subjective experience will depend heavily on the *environment* in which the drug is taken. The experience of drug taking in the company of friends in pleasant surroundings is likely to be completely different from that elicited by the same dose of the drug administered to volunteer subjects studied under laboratory conditions or, as in some of the earlier American studies, to convicts in prison who had "volunteered" as experimental subjects.

Third, the drug experience will depend on the *mood and personality* of the user, their *familiarity* with cannabis, and their *expectations* of the drug. The same person may experience entirely different responses to the drug depending on whether he or she is depressed or elated beforehand. Familiarity with the drug means that the user knows what to expect, whereas the inexperienced user may find some of the elements of the drug experience unfamiliar and frightening. The person using the drug for medical reasons has entirely different expectations from those of the recreational user and commonly finds the intoxicating effects of cannabis disquieting and unpleasant.

There are many detailed descriptions of the marijuana experience in the literature. Among the best known are the flowery and often lurid literary accounts of the 19th-century French authors Baudelaire, Gautier, and Dumas and those written by the 19th-century Americans Taylor and Ludlow. Ludlow's (1857) book, *The Hasheesh Eater*, gives one of the best accounts and is quoted frequently here. Fitz Hugh Ludlow was an intelligent young man who experimented with various mind-altering drugs. He first encountered marijuana at the age of 16 years in the local pharmacy, and he became fascinated by the drug and eventually addicted to it. His book vividly describes the cannabis experience, although it is worth bearing in mind that he regularly consumed doses of herbal cannabis extract that would be considered very large by current standards— probably equivalent to several cannabis cigarettes in one session. In modern times, there have been several surveys of the experiences of marijuana users. Among these, the book by E. Goode (1970), *The Marijuana Smokers*, and that by J. Berke and C. H. Hernton (1974), *The Cannabis Experience*, which review the experiences of young American and British cannabis users in the 1960s and 1970s, respectively, are particularly useful. For "trip reports" from contemporary marijuana users, see the Erowid experience vaults at https://erowid.org/experiences/subs/exp_Cannabis.shtml.

The various stages of the experience can be separated into the *buzz*, which leads to the *high* and then the *stoned* states, and finally the *come-down*. The buzz is a transient stage, which may arrive fairly quickly when smoking. It is a tingling sensation felt in the body, in the head, and often in the arms and legs, accompanied by a feeling of dizziness or light-headedness:

> With hashish a "buzz" is caused, i.e. a tingling sensation forms in the head and spreads through the neck and across the shoulders. With a very powerful joint this sensation is sometimes "echoed" in the legs.
>
> "Usually the first puff doesn't affect me, but the second brings a slight feeling of dizziness and I get a real 'buzz' on the third. By

this I mean a sudden wave of something akin to dizziness hits me. It's difficult to describe. The best idea I can give is to say that for a moment the whole room, people, and sounds around me recede into the distance and I feel as I my mind contracted for an instant. When it has passed I feel 'normal' but a bit 'airy-fairy.'" (Berke and Hernton, 1974)

During the initial phase of intoxication, the user will often experience bodily sensations of warmth (caused by the drug-induced relaxation of blood vessels and increased blood flow, for example, to the skin). The increase in heart rate caused by the drug may also be perceived as a pounding pulse. Marijuana smokers also commonly feel a dryness of the mouth and throat and may become very thirsty. This may be exacerbated by the irritant effects of marijuana smoke, but it is also experienced when the drug is taken by mouth.

The influence of the drug on the mind is far-reaching and varied; the marijuana high is a very complex experience. It is only possible to highlight some of the common features here. THC has profound effects on the highest centers in the brain and alters both the manner in which sensory inputs are normally processed and analyzed and the thinking process itself. Mental and physical excitement and stimulation usually accompany the initial stages of the "high." The drug is a powerful euphoriant, as described so well by Ludlow (1857). Some hours after taking an extract of cannabis, he was

> smitten by the hashish thrill as by a thunderbolt. Though I had felt it but once in life before, its sign was as unmistakable as the most familiar thing of daily life. ... The nearest resemblance to the feeling is that contained in our idea of the instantaneous separation of soul and body.

The hashish high was experienced while Ludlow was walking with a friend, and the effects could be felt during the walk and after they returned home:

The road along which we walked began slowly to lengthen. The hill over which it disappeared, at the distance of half a mile from me, soon became to be perceived as the boundary of the continent itself. . . . My awakened perceptions drank in this beauty until all sense of fear was banished, and every vein ran flooded with the very wine of delight. Mystery enwrapped me still, but it was the mystery of one who walks in Paradise for the first time. . . . I had no remembrance of having taken hasheesh. The past was the property of another life, and I supposed that all the world was revelling in the same ecstasy as myself. I cast off all restraint; I leaped into the air; I clapped my hands and shouted for joy. . . . I glowed like a new-born soul. The well known landscape lost all of its familiarity, and I was setting out upon a journey of years through heavenly territories, which it had been the longing of my previous lifetime to behold. . . . In my present state of enlarged perception, time had no kaleidoscope for me; nothing grew faint, nothing shifted, nothing changed except my ecstasy, which heightened through interminable degrees to behold the same rose-radiance lighting us up along our immense journey. . . . I went on my way quietly until we again began to be surrounded by the houses of the town. Here the phenomenon of the dual existence once more presented itself. One part of me awoke, while the other continued in perfect hallucination. The awakened portion felt the necessity of keeping inside on the way home, lest some untimely burst of ecstasy should startle more frequented thoroughfares.

The 19th-century physician H. C. Wood of Philadelphia described his experimental use of cannabis extract as follows:

It was not a sensuous feeling, in the ordinary meaning of the term. It did not come from without; it was not connected with any passion or sense. It was simply a feeling of inner joyousness; the heart seemed buoyant beyond all trouble; the whole system felt as though all sense of fatigue were forever banished; the mind gladly ran riot, free constantly to leap from one idea to another, apparently unbound from its ordinary laws. I was disposed to laugh; to make comic gestures. (as cited in Walton, 1938, p. 88)

The initial stages of intoxication are accompanied by a quickening of mental associations, and this is reflected typically by a sharpened sense of humor. The most ordinary objects or ideas can become the subjects of fun and amusement, often accompanied by uncontrollable giggling or laughter:

"I often feel very giggly, jokes become even funnier, people's faces become funny and I can laugh with someone else who's stoned just by looking at them."

"I would start telling long involved jokes, but would burst out laughing before completion."

"I nearly always start laughing when in company and have on numerous occasions been helpless with laughter for up to half-an-hour non-stop." (Berke and Hernton, 1974)

This effect of the drug is difficult to explain because so little is known about the brain mechanisms involved. Humor and laughter seem to be unique human features. A sharpened sense of humor and increased propensity to laugh are not unique to THC; they are seen with other intoxicants, notably alcohol. A visit to any lively pub in Britain will confirm this phenomenon. However, THC does seem to be remarkably powerful in inducing a state that has been described as "fatuous euphoria."

As the level of intoxication progresses from "high" to "stoned" (if the dose is sufficiently large), users report feeling relaxed, peaceful, and calm; their senses are heightened and often distorted; they may have apparently profound thoughts; and they experience a curious change in their subjective sense of time. As in a dream, the user believes that far more time has passed than it has in reality. As Goode (1970) states,

Somehow, the drug is attributed with the power to crowd more "seeming" activity into a short period of time. Often nothing will appear to be happening to the outside observer, aside from a few individuals slowly smoking marijuana, staring into space and, occasionally, giggling at nothing in particular, yet each mind will be crowded with past or imagined events and emotions, and significance of massive proportions will be attributed to the scene, so that activity will be imagined where there is none. Each minute will be imputed with greater significance; a great deal will be thought to have occurred in a short space of time. More time will be conceived of as having taken place. Time, therefore, will be seen as being more drawn out.

Young British cannabis users report similar experiences:

"The strongest feeling I get when I am most stoned is a very confused sense of time. I can start walking across the room and become blank until reaching the other side, and when I think back it seems to have taken hours. Many records seem to last much longer than they should."

"Perhaps the 'oddest' experience is the confusion of time. One could walk for five minutes and get hung up on something and think that it is an hour later or the other way around, i.e. watch a movie and think it only took five minutes instead of two hours." (Berke and Hernton, 1974)

Research work at Stanford University in the 1970s by Frederick Melges and colleagues on cannabis users led them to conclude that the disorientation of time sense might represent a key action of the drug, from which many other effects flowed (Melges et al., 1971). Their subjects tended to focus on the present to the exclusion of the past or future. Not having a sense of past or future could lead to the sense of depersonalization that many users experience. A focus on the present might also account for a sense of heightened perception, by isolating current experiences from those in the past. This loss of the normal sense of time is probably related to the rush of ideas and sensations experienced during the marijuana high. The user will become unable to maintain a continuous train of thought and no longer able to hold a conversation:

> "Sometimes I find it difficult to speak simply because I have so many thoughts on so many different things that I can't get it all out at once." (Berke and Hernton, 1974)

Perception becomes more sensitive, and the user has a heightened appreciation of everyday experiences. A nurse describes seeing the Chinese-style pagoda in Kew Gardens in London under the influence of marijuana:

> "It was like the pagoda had been painted a bright red since I had last seen it—about an hour before. The colour was not just bright, but more than bright, it was a different hue altogether, a deep red, with lots of added pigments, a red that was redder than red. It was a red that leapt out at you, that scintillated and pulsated amid the grey sky of a typical dull English afternoon. Never in a thousand years will I forget that sight. It was like my eyes had opened to colour for the first time. And ever since then, I have been able to appreciate colour more deeply." (Berke and Hernton, 1974)

New insight and appreciation of works of art have often been reported. Many users report that their appreciation and enjoyment of music are especially enhanced while high; they gain the ability to comprehend the structure of a piece of music, the phrasing, tonalities, and harmonies and the way that they interact. Some musicians believe that their performance is enhanced by marijuana, and this undoubtedly accounted for the popularity of marijuana among jazz players in the United States in the early years of the century. Ludlow (1857) described his experience of attending a concert while under the influence of the drug:

A most singular phenomenon occurred while I was intently lis-
tening to the orchestra. Singular, because it seems one of the most
striking illustrations I have ever known of the preternatural activity
of sense in the hasheesh state, and in an analytic direction. Seated
side by side in the middle of the orchestra played two violinists.
That they were playing the same part was obvious from their perfect uniformity in bowing; their bows, through the whole piece,
rose and fell simultaneously, keeping exactly parallel. A chorus of
wind and stringed instruments pealed on both sides of them, and
the symphony was as perfect as possible; yet, amid all that harmo-
nious blending, I was able to detect which note came from one vi-
olin and which from the other as distinctly as if the violinists had
been playing at the distance of a hundred feet apart, and with no
other instruments discoursing near them.

Although there is no evidence that cannabis is an aphrodisiac,
it may enhance the pleasure of sex for some people because of their
heightened sensitivity and loss of inhibitions. But if the user is not
in the mood for sex, getting high by itself will not alter that: "Hash
increases desire when desire is already there, but doesn't create desire out of nothing."

The increased sensitivity to visual inputs tends to make marijuana users favor dimly lit rooms or dark sunshades because they
find bright light unpleasant. The mechanisms in the brain that modulate and filter sensory inputs and set the level of sensitivity clearly
become disinhibited. The analysis of sensory inputs by the cerebral
cortex also changes, in some ways becoming freer ranging—in other
ways becoming less efficient. For example, as intoxication becomes
more intense, sensory modalities may overlap so that, for example,
sounds are seen as colors and colors contain music—a phenomenon psychologists refer to as "synesthesia":

"I have experienced synesthesia—I 'saw' the music from an Indian
sitar LP. It came in the form of whirling mosaic patterns. I could

change the colours at will. At one time a usual facet of a high was
that musical sound would take on a transparent crystal, cathedral,
spatial quality." (Berke and Hernton, 1974)

The peak of intoxication may be associated with hallucinations—
that is, seeing and hearing things that are not there. Cannabis does
not induce the powerful visual hallucinations that characterize the
drug LSD, but fleeting hallucinations can occur, usually in the visual
domain:

> ". . . occasionally hallucinations. I will see someone who is not
> there, the much described 'insects' which flutter around at the edge
> of vision, patterns move and swirl." (Berke and Hernton, 1974)

At the most intense period of the intoxication, the user finds
difficulty in interacting with others and tends to withdraw into an
introspective state. Thoughts tend to dwell on metaphysical or phil-
osophical topics, and the user may experience apparently transcen-
dental insights:

> "The final stage I entered was concurrent with the Buddhist en-
> lightenment, at least according to my admittedly limited under-
> standing of it. My mind ceased to be a distinct entity, but became
> One with the entire universe. At the same time, however, Nothing
> existed at all. It was infinity and zero at the same time. All was one
> and all one was all. But all was also nothing. Continuing with the
> stick figure analogy above, my old stick figure self had moved from
> merely conceiving of a third dimension to existing fully in it and
> eventually entering a black hole into an indescribable dimension-
> less realm above this. I was not familiar with the concept of ego
> death at the time of this experience, but I now believe what I expe-
> rienced is very similar to this. I have read accounts where ego death
> is a terrifying experience, as the ego has trouble letting go and the
> subject confuses ego death with physical death. In my experience,

however, there was absolutely no struggle where the ego attempted to hold on. It simply let go as if being slowly washed and dissolved away. It was completely natural. There was no concept of whether or not I 'wanted' this to happen, it simply happened. If a feeling must be attached to it, that feeling I would say is sublime peace and oneness. Perhaps my lack of experience with psychedelics has fooled me into thinking I experienced ego death when it really wasn't." (https://erowid.org/experiences/exp.php?ID=86671; accessed April 21, 2017)

The peak period of intoxication is also commonly associated with daydreams and fantasies:

> Fantasies, your thoughts seem to run along on their own to the extent that you can relax and "watch" them (rather like an intense day-dream). . . . Images come to mind that may be funny, curious, interesting in a story-telling sort of way, or sometimes horrific (according to mood). Also many other variations. (Berke and Hernton, 1974)

The nature of the fantasies varies according to personality and mood. One of the most common fantasies is that of power. The user feels that he is a god, a superman; that he is indestructible; and that all his desires can be satisfied immediately. Not surprisingly, people find such fantasy states enjoyable and cite them as one of the reasons for their continued use of the drug. Ludlow (1857) described it as follows:

> My powers became superhuman; my knowledge covered the universe; my scope of sight was infinite. . . . All strange things in mind, which had before been my perplexity, were explained—all vexed questions solved. The springs of suffering and of joy, the action of the human will, memory, every complex fact of being, stood forth

before me in a clarity of revealing which would have been the sublimity of happiness.

A curious feature of the cannabis high is that its intensity may vary intermittently during the period of intoxication, with periods of lucidity intervening. There is often the strange feeling of "double consciousness." Subjects speak of watching themselves undergo the drug-induced delirium—of being conscious of the condition of their intoxication yet being unable or unwilling to return to a state of normality. Experienced users can train themselves to act normally and may even go to work while intoxicated.

As the effects of the drug gradually wear off, there is the coming down phase. This may be preceded by a sudden feeling of hunger ("munchies"), often associated with feelings of "emptiness" in the stomach. There is a particular craving for sweet foods and drinks, and there is an enhanced appreciation and enjoyment of food:

> "When I am coming down I generally feel listless and physically weak. . . . Often the high ends with a feeling of tiredness, this can be overcome, but is usually succumbed to when possible if not by sleep, by a long lay down. . . . Conversation initially becomes lively and more intense but as the high wears off and everyone becomes sleepy it usually stops." (Berke and Hernton, 1974)

The cannabis high is often followed by sleep, sometimes with colorful dreams.

However, the cannabis experience is not always pleasant. Inexperienced users in particular may experience unpleasant physical reactions. Nausea is not uncommon, and it may be accompanied by vomiting, dizziness, and headache. As users become more experienced, they learn to anticipate the wave of light-headedness and dizziness that are part of the "buzz." Even regular users will sometimes have very unpleasant experiences, particularly if they take a larger dose of drug than normal. The reaction is one of intense fear and anxiety, with symptoms resembling those of a panic attack and sometimes accompanied by physical signs of pallor (the so-called "whitey"), sweating, and shortness of breath. The psychic distress can be intense, as described by young British users:

> "I once had what is known as 'the horrors' when I had not been smoking long. The marijuana was a very strong variety, far stronger than anything I had ever smoked before, and I was in an extremely tense and unhappy personal situation. I lost all sense of time and place and had slight hallucinations—the walls came and went, objects and sounds were unreal and people looked like monsters. It was hard to breathe and I thought I was going to die and that no one would care."
>
> "I have felt mentally ill twice when using hashish. On both occasions I felt that I could control no thoughts whatsoever that passed through my mind. It was as though my brain had burst and was distributed around the room. I knew that a short time before-hand I had been quite sane, but that now I was insane and I was desperate because I thought that I would never reach normality again. I saw myself in the mirror, and although I knew that it, the person I saw was me, she appeared to be a complete stranger, and I realized that this was how others must see me. Then the head became estranged from the body—flat piece of cardboard floating a few inches above the shoulders. I was completely horrified, but fascinated, and stood and watched for what must have been some minutes." (Berke and Hernton, 1974)

Ludlow's (1857) description of a cannabis-induced horror is particularly graphic. After he had taken a much larger dose of cannabis than usual—in the mistaken belief that the preparation was weaker than the one he had used most recently—he went to sleep in a dark room:

> I awoke suddenly to find myself in a realm of the most perfect clarity of view, yet terrible with an infinitude of demoniac shadows.

Perhaps, I thought, I am still dreaming; but no effort could arouse me from my vision, and I realized that I was wide awake. Yet it was an awaking which, for torture, had no parallel in all the stupendous domain of sleeping incubus. Beside my bed in the centre of the room stood a bier, from whose corners drooped the folds of a heavy pall; outstretched upon it lay in state a most fearful corpse, whose livid face was distorted with the pangs of assassination. The traces of a great agony were frozen into fixedness in the tense position of every muscle, and the nails of the dead man's fingers pierced his palms with the desperate clinch of one who has yielded not without agonizing resistance. . . . I pressed my hands upon my eyeballs till they ached, in intensity of desire to shut out this spectacle; I buried my head in the pillow, that I might not hear that awful laugh of diabolic sarcasm. . . . The stony eyes stared up into my own, and again the maddening peal of fiendish laughter rang close beside my ear. Now I was touched upon all sides by the walls of the terrible press; there came a heavy crush, and I felt all sense blotted out in the darkness.

I awaked at last; the corpse had gone, but I had taken his place upon the bier. In the same attitude which he had kept I lay motionless, conscious, although in darkness, that I wore upon my face the counterpart of his look of agony. The room had grown into a gigantic hall, whose roof was framed of iron arches; the pavement, the walls, the cornice were all of iron. The spiritual essence of the metal seemed to be a combination of cruelty and despair. . . . I suffered from the vision of that iron as from the presence of a giant assassin.

But my senses opened slowly to the perception of still worse presences. By my side there gradually emerged from the sulphurous twilight which bathed the room the most horrible form which the soul could look upon unshattered—a fiend also of iron, white hot and dazzling with the glory of the nether penetralia. A face that was the ferreous incarnation of all imaginations of malice and irony looked on me with a glare, withering from its intense heat, but still more from the unconceived degree of inner wickedness which it symbolized. . . . Beside him another demon, his very twin, was rocking a tremendous cradle framed of bars of iron like all things else, and candescent with as fierce a heat as the fiends.

And now, in a chant of the most terrible blasphemy which it is possible to imagine, or rather of blasphemy so fearful that no human thought has ever conceived it, both the demons broke forth, until I grew intensely wicked merely by hearing it . . . suddenly the nearest fiend, snatching up a pitchfork (also of white hot iron), thrust it into my writhing side, and hurled me shrieking into the fiery cradle.

After more terrible visions, Ludlow eventually cried out for help and a friend brought him water and a lamp, upon which his terrors ceased. He was to experience both "superhuman joy and super-human misery" from the drug, but he became dependent on it and took it for many years until, after a long struggle, he finally gave it up.

Subjective Effects of Synthetic Cannabinoids

The psychoactive effects of synthetic cannabinoids can vary considerably. Effects are dependent on the usual context issues and the potency of the product. Some brands produce a "cannabis-like" dreamy euphoria and "stoned" feeling, with elevated mood, relaxation, and altered perception, whereas the more potent brands elicit cannabis-like effects in addition to a range of "un-cannabis like" effects. These may include effects similar to those elicited by dissociative anesthetics (e.g., ketamine) and/or hallucinations such as those elicited by psychedelic drugs (e.g., LSD). These effects are often accompanied by unpleasant side effects, including symptoms of psychosis, delusional and disordered thinking, detachment from reality, and paranoia. Other negative effects include agitation, panic, anxiety, and dysphonia. The range of subjective effects elicited by the more potent synthetic cannabinoids is greater than those elicited by cannabis, and the likelihood of experiencing unpleasant side effects is greater. In a literature review, it was found that the effects of syuthetic canabimnoids were reportedly similar to those of marijuana and were well tolerated. Not surprisingly, the individuals all reported that synthetic cannabinoid use effectively alleviated cannabis withdrawal. Potential toxicity, such as acute anxiety and psychosis, was also observed (Spaderna and D'Souza, 2011). Drug-induced death is not uncommon after use of synthetic cannabinoids, whereas death from overdose of cannabis is rare (see Chapter 6). (For reports of experiences of young people taking synthetic cannabinoids, see Erowid experience vaults at https://erowid.org/experiences/subs/exp_Cannabis.shtml; click on "Cannabinoid Receptor Agonists.")

Comparisons of Marijuana with Alcohol

Alcohol and marijuana are both drugs usually taken in a social context for recreational purposes. Alcohol could be described as the intoxicant for the older generation and marijuana that for the young, although both drugs are quite often consumed together. How do they compare in their effects on the brain? In many ways, they are quite similar. A number of studies performed under laboratory conditions have reported that users find it difficult to distinguish between the immediate subjective effects of acute intoxication with the two drugs. Like marijuana, alcohol causes psychomotor impairments, a loss of balance, and a feeling of dizziness or lightheadedness. In terms of cognitive performance, both drugs cause impairments in short-term memory while leaving the recall of long-term memories intact. But there are obviously some notable differences. Interestingly, the sense of time perception in subjects intoxicated with alcohol is changed in the opposite direction to that observed with marijuana. Tests similar to those described previously for marijuana reveal that whereas marijuana "speeds up" the internal clock, alcohol slows it down—one minute may seem like an infinity to the marijuana user but feels like only 30 seconds to the alcohol user. In terms of their effects on driving, while studies to determine whether cannabis causes an increased risk of accidents have been inconclusive, there is unanimous agreement that alcohol use increases the risk of a crash. Furthermore, the risk from driving under the influence of both alcohol and cannabis is greater than the risk of driving under the influence of either alone (Sewell et al., 2009). Whereas marijuana tends to make users relaxed and tranquil, alcohol may release aggressive and violent behavior. In terms of the long-term effects of chronic use, alcohol has none of the subtlety of marijuana. Heavy long-term use can lead to organic brain damage and psychosis or dementia (a condition known as Korsakoff's syndrome), and even moderately heavy use can lead to quite severe persistent intellectual impairment.

What Can Animal Behavior Experiments Tell Us?

Discriminative Stimulus Effects

Studying the actions of psychotropic drugs in animals is inherently difficult—the animals cannot tell us what they are experiencing. The application of ingenious behavioral tests, however, can tell us a great deal about how a drug "feels" to an animal. One technique that is widely used assesses the "discriminative stimulus effects" of CNS drugs. In this test, the animals, usually rats, are trained to press a lever in their cage in order to obtain a food reward, usually a small attractively flavored food pellet, and the reward is given automatically after a certain number of lever presses. The animals are then presented with two alternative levers and must learn to press one (the saline lever) if they had received a saline injection just before the test session or the other (the drug lever) if they had been injected with the active test drug. Pressing the "wrong" lever provides no food reward. In other words, the animal is being asked, "How do you feel? Can you tell that you just received a psychoactive drug?" Animals are tested every day for several weeks, receiving drug or saline randomly, and they gradually learn to discriminate the active drug from the placebo (saline). They are judged to have learned the discrimination if they successfully gain a food reward with a minimal number of presses of the "wrong" lever. This technique has provided a great deal of valuable information about cannabis and related drugs. The method has been applied to phytocannabinoids, endocannabinoids, and novel synthetic cannabinoids, and it can inform which substances are psychoactive and provide valuable information on their potency (Wiley et al., 2016). Rats and monkeys successfully recognize THC or various synthetic cannabinoids within 2 or 3 weeks of daily training (Figure 3.2).

The doses of cannabinoids that animals recognize are quite small—less than 1 mg/kg orally for THC and much less for the

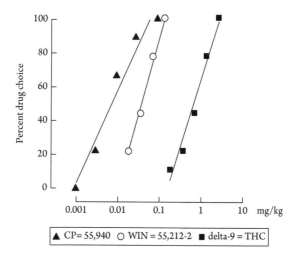

FIGURE 3.2 Rats trained to discriminate an injection of the synthetic cannabinoid WIN-55,212–2 (0.3 mg/kg, given subcutaneously) from saline also recognize lower doses of this compound and the other psychoactive cannabinoids CP-55,940 (given subcutaneously) and THC (given orally). The graph shows the percentage of animals selecting the "drug" lever after various doses of the cannabinoids. Results are from a group of nine rats.

Source: From Pério et al. (1996).

synthetic cannabinoids WIN-55,212-2 and CP-55,940 given sub-cutaneously (0.032 and 0.007 mg/kg, respectively; see Figure 3.2) (Torbjörn et al., 1974; Wiley et al., 1995; Pério et al., 1996). These doses are in the range known to cause intoxication in human subjects. When animals have been trained to discriminate one of these drugs, the experimenter can substitute a second or third drug and ask the animal another question: "Can you tell the difference between this drug and the one you were previously trained to recognize?" The results of such experiments show that rats and monkeys trained to recognize one of the cannabinoids will generalize (i.e., judge to be the same) to any of the others. They will not generalize, however, to a variety of other CNS-active drugs, including psilocybin, morphine, benzodiazepines, or phencyclidine, suggesting that cannabinoids produce a unique spectrum of CNS effects that the animal can recognize. In all of these studies, it was found that rimonabant completely blocked the effects of the cannabinoids— that is, when animals are treated with the cannabinoid together with the antagonist, they are no longer able to recognize the cannabinoid. These results thus provide further strong support for the hypothesis that the CNS effects of THC and other cannabinoids are directly attributable to their actions on the CB-1 receptor in the brain.

Using these techniques, one can also ask whether the endogenous cannabinoid anandamide really mimics THC and the other cannabinoids. Rats trained to recognize a synthetic cannabinoid do generalize to anandamide, but high doses of anandamide are needed because it is rapidly inactivated in the body. Monkeys do not generalize to anandamide, probably because it is inactivated too quickly. However, if monkeys are given a synthetic derivative of anandamide that is protected against metabolic inactivation, then they will generalize to this.

In another study, rats were trained to recognize THC and were then exposed to cannabis resin smoke. They recognized the cannabis smoke as though it were THC and showed full generalization. In the same study, it was found that Δ^9-THC and Δ^8-THC were recognized interchangeably, but there was no generalization between cannabinol or cannabidiol and either THC or cannabis smoke. These results support the hypothesis that THC is the major psychoactive component in cannabis resin and suggest that cannabinol and cannabidiol have little psychoactive effects (Jarbe and McMillan, 1980).

Effects on Cognition

There is a large literature on the effects of THC and other cannabinoids on various aspects of animal behavior (Adams and Martin, 1996; Bab and Alexander, 2011). Unfortunately, many studies have used very high doses of THC, and the results consequently may have little relevance to how the drug affects the human brain. The human intoxicant dose for THC is less than 0.1 mg/kg, but doses several hundred times higher have often been used in animal studies. Such high doses of THC depress most aspects of animal behavior and may cause catalepsy and eventually sleep. Work with much smaller doses of cannabinoids has shown the importance of using the appropriate dose. The acute administration of low doses of cannabinoid agonists to laboratory rodents produces deficits in working memory, attentional function, and reversal learning, and animal studies have helped pinpoint the prefrontal cortex as a key site for the neurochemical processes underlying these cognitive deficits (Egerton et al., 2006). Further animal and human research has shown the importance of corticolimbic circuits in mediating the cognitive and emotional responses to cannabinoids, including endocannabinoids. The prefrontal cortex, amygdala, and ventral hippocampus are proposed as key brain regions (Silveira et al., 2016).

There are a number of ways of assessing working memory in animals. One model frequently used in rodents to assess spatial working memory is the radial maze. In this model, a rat or mouse is placed at the center of a maze that has eight arms projecting away from the central area. At the start of each experiment, all eight arms contain a food reward. To begin, the animal is placed at the center of the maze and enters one arm to retrieve a food reward. The animal is then returned to the central area, and all eight arms are temporarily blocked by sliding doors. After a delay, usually of only a few seconds, the doors are opened again, and the animal is free to retrieve more food rewards. Success depends on being able to remember which arms have already been visited to avoid fruitless quests. After daily training for 2 or 3 weeks, the animals become quite expert at the task and retrieve all eight food rewards while making few errors. THC and other cannabinoids will disrupt the behavior of such trained animals in a dose-dependent manner. Furthermore, this effect of the cannabinoids can be prevented by rimonabant, showing that it is due to an action of THC on CB-1 receptors. The synthetic cannabinoids CP-55,940 and WIN-55,212-02 are also effective in this model, and they are considerably more potent than THC. Another behavioral test of memory that can be employed both in rodents and in monkeys is the "delayed matching to sample" task (Lind et al., 2015). When using this test in monkeys, an animal is confronted with a number of alternative panels on a touch screen. At the start of the experiment, one of these panels is illuminated and the screen then goes dead, preventing the animal from making any immediate response. After a delay, usually of 30–90 seconds, all the panels on the screen are illuminated, and the animal has to remember which one was illuminated earlier and press it to obtain a food reward. After daily training sessions, animals become proficient at this spatial memory task and make few errors. THC and other cannabinoids again disrupt behavior in these tests of working memory. Using a variant of this task, similar results have been observed in rats (Heyser et al., 2016).

The ongoing release of endogenous cannabinoids in brain may play a role in modulating memory processes. When adult rats or mice are exposed for the first time to a juvenile animal, they spend some time contacting and investigating it. If the adult is exposed to the same juvenile within 1 hour of the first encounter, it appears to recognize that it has already encountered this juvenile and will spend less time investigating it. If the delay between trials is increased to 2 hours, however, the adult seems to have largely forgotten the original encounter and investigates the juvenile animal thoroughly once more. This short-term memory of "social recognition" appears to rely mainly on olfactory cues. Animals treated with low doses of the antagonist rimonabant showed improved memory function in this test and were able to retain the social recognition cues for 2 hours or more. The powerful effects of cannabinoid drugs in this test may be related to the fact that social recognition in rodents importantly involves olfactory cues, and the CB-1 receptor is present in especially high densities in the olfactory regions of the brain.

The cellular basis of these effects of cannabis on higher brain function remains unclear. Cannabinoids have been shown to disrupt the phenomenon of "long-term potentiation" in the hippocampus (Riedel and Davies, 2005). In this model, slices of rat hippocampus are incubated in saline, and electrical activity is recorded from nerve cells by miniature electrodes. A burst of electrical stimulation to the input nerve pathways to the hippocampus leads to a long-lasting potentiation of synaptic transmission in this circuit so that further periods of less intense stimulation lead to greater responses than previously. This form of plasticity in neural circuits is thought to be critical in the laying down of memory circuits in the brain.

Anti-Anxiety Effects

Using tests that normally inhibit rodent behavior (exposure to open field or elevated plus maze), studies have shown that cannabinoid agonists produce behavioral effects in rats suggestive of an anti-anxiety effect (De Fonseca et al., 1997; Bruijnzeel et al., 2016; Di et al., 2016; Patel, 2017). A role for marijuana in the treatment of anxiety is one of the targets of medical marijuana (see Chapter 5).

Chapter 4

Endocannabinoids

For a review of endocannabinoids, see Pertwee (2015b).

Discovery

The existence of specific receptors for cannabinoids in brain and in other tissues suggested that they were there for some reason. The receptors had not evolved simply to recognize a psychoactive drug derived from a plant, just as the opiate receptor was not in the brain simply to recognize morphine. In the 1970s, the discovery of the opiate receptor in brain prompted an intense search for the naturally occurring brain chemicals that might normally activate this receptor, and this revealed the existence of a family of brain peptides known as the "endorphins" (*endo*genous mor*phin*es). Similarly, the discovery of the cannabis receptor prompted a search for the naturally occurring cannabinoids (now known as "endocannabinoids") (for reviews, see Axelrod and Felder, 1998; Felder and Glass, 1998; Mechoulam et al., 1998; Di Marzo et al., 2005). These discoveries have radically changed the way in which scientists view this field of research. It has changed from a pharmacological study of how the plant-derived psychoactive drug delta-9-tetrahyrocannabinol (THC) works in the brain to a much broader field of biological research on a unique natural control system, now often referred to as the "cannabinoid system." The term *cannabinoid*, originally used to describe the 21-carbon substances found in cannabis plant extracts, is now used to define any compound that is specifically recognized by cannabinoid receptors. Anandamide is a fairly simple chemical, and could readily be synthesized in larger quantities by chemists (Figure 4.1). Many studies have been performed on this endocannabinoid, confirming that it has essentially all of the pharmacological and behavioral actions of THC in various animal models—including the "Billy Martin tetrad"—although when given to animals it is considerably less potent than THC because it is rapidly inactivated. The discovery of anandamide was not the end of the story. Mechoulam and colleagues went on to identify a second naturally occurring cannabinoid, also a derivative of arachidonic acid, known as 2-arachidonylglycerol (2-AG) (Hanus et al., 2001) (see Figure 4.1). This, too, was synthesized and proved to have THC-like actions in various biological tests, including whole animal behavioral models, and potency similar to that of anandamide. Since then, another 12 endocannabinoids have been described (see Figure 4.1). The 13 endocannabinoids currently known are anandamide, 2-AG, noladin ether, dihomo-γ-linolenoylethanolamide, virodhamine, oleamide, docosahexaenoylethanolamide, eicosapentaenoylethanolamide, sphingosine, docosatetraenoylethanolamide, N-arachidonoyldopamine (NADA), N-oleoyldopamine, and hemopressin (Pertwee, 2015b).

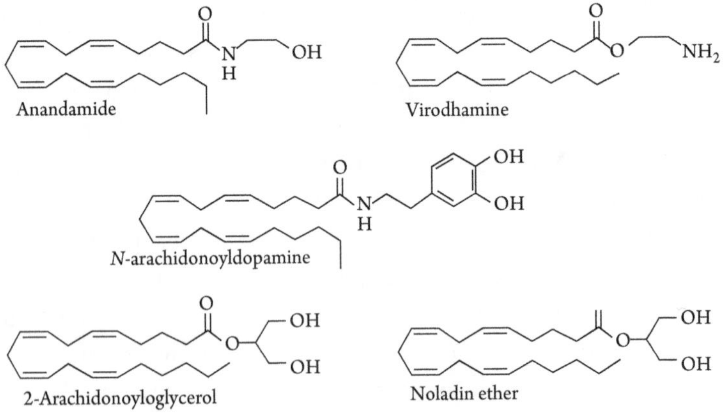

FIGURE 4.1 Naturally occurring endocannabinoids.

In vitro findings suggest that the first 8 of these compounds can activate CB-1 and sometimes also CB-2 receptors and that another 2 of these compounds are CB-1 receptor antagonists (sphingosine) or antagonists/inverse agonists (hemopressin). There is also evidence for the existence of at least 3 allosteric endocannabinoids that appear to target allosteric sites on cannabinoid receptors in vitro, either as negative allosteric modulators of the CB-1 receptor (pepcan-12 and pregnenolone) or as positive allosteric modulators of this receptor (lipoxin A_4) or of the CB-2 receptor (pepcan-12S) (Howlett et al., 2004; Pertwee, 2015b).

The endocannabinoids are part of a large family of other lipid signaling molecules derived from arachidonic acid that includes the prostaglandins and leukotrienes, important mediators of inflammation. Far less is known about the newer members of the endocannabinoid group, and it remains unclear whether they all play important functional roles. Study of the endocannabinoids has become a very active field of both basic and applied research; as of July 2017, PubMed listed more than 6,000 scientific publications involving the endocannabinoids. Although this is an exciting new area of biology, it would be impossible to provide a detailed survey here without a very lengthy chapter. This monograph is focused primarily on what is known about cannabis and the advantages and disadvantages of its medical and recreational use; thus, endocannabinoids are reviewed only briefly.

Biosynthesis and Inactivation of Endocannabinoids

For a review of the biosynthesis and inactivation of endocannabinoids, see Di Marzo et al. (2005).

Many of the endocannabinoids are derived from the unsaturated fatty acid arachidonic acid, which is one of the fatty acids found commonly in the lipids in all cell membranes. Anandamide is synthesized by the enzyme phospholipase D, whereas 2-AG is synthesized by a different route involving an enzyme known as DAG-lipase. Biosynthetic routes for the endocannabinoids derived from ethanolamides remain to be elucidated. Inhibitors of the biosynthetic enzyme offer an alternative to cannabinoid receptor antagonists in dampening cannabinoid activity, and there has been some progress in discovering such inhibitors (Pertwee, 2015b).

As with the prostaglandins and leukotrienes, the endocannabinoids are not stored in cells awaiting release but, rather, are synthesized on demand. The stimulus that triggers biosynthesis is a sudden influx of calcium on activation of the cell. The rate of biosynthesis of anandamide and 2-AG in brain is increased, for example, when nerve cells are activated by exposure to the excitatory amino acid L-glutamate. The dogma that endocannabinoids are exclusively synthesized and released on demand has been challenged by the proposal that intracellular trafficking of anandamide to intracellular stores (adiposomes) and binding proteins may be important (Maccarrone et al., 2010).

As with other biological messenger molecules, the endocannabinoids are rapidly inactivated after their formation and release. Both anandamide and 2-AG are broken down by hydrolytic enzymes. The enzymes fatty acid amide hydrolase (FAAH) and monoacylglycerol lipase (MGL) seem to play a key role. FAAH has been cloned and sequenced; it belongs to the large family of serine hydrolytic enzymes. MGL is a membrane-associated member of the serine hydrolase superfamily. FAAH and MGL co-localize with CB-1 receptors in the hippocampus, cerebellum, and amygdala but are distributed in different areas of the synapse: FAAH is expressed postsynaptically, whereas MGL is expressed in the presynaptic terminal. Both enzymes play an important role in retrograde endocannabinoid signaling. Hydrolysis of these endocannabinoids by FAAH and MGL prevents persistent CB-1 receptor activation and desensitization. A number of inhibitors of both FAAH and MGL have been developed, and these have potential medical uses in prolonging the actions of released endocannabinoids (discussed later).

Whether a specific transport protein exists to shuttle anandamide and other endocannabinoids into cells where they can then be metabolized has proved controversial. Currently, most scientists seem to accept that there is no need for an "endocannabinoid transporter" mechanism for small uncharged lipophilic molecules and that endocannabinoids simply diffuse into cells and are rapidly degraded by FAAH, which acts as a "sink" drawing them in (Fowler, 2013; Nicolussi and Deutsch, 2015). However, some argue that anandamide uptake is real because it can still be demonstrated in the presence of inhibitors of FAAH or in FAAH knockout mice (Khairy and Houssen, 2010).

Physiological Functions of Endocannabinoids

Retrograde Signaling Molecules at Synapses

For reviews of retrograde signaling molecules, see Howlett et al. (2004) and Kano et al. (2009).

Elphick and Egertová (2001) undertook immunohistochemical mapping studies of the regional distribution of CB-1 receptors and the enzyme FAAH in brain and found that there was considerable overlap, suggesting a complementary relationship between the two at the synaptic level. They postulated the existence of a retrograde cannabinoid signaling mechanism whereby endogenous cannabinoids released in response to synaptic activation feed back to presynaptic receptors on these axon terminals and are subsequently inactivated by FAAH after their uptake into the postsynaptic compartment. This hypothesis has been supported independently by neurophysiological findings.

A phenomenon known as depolarization-induced suppression of inhibition (DSI) has been known to neurophysiologists for some years. It is a form of fast retrograde signaling from postsynaptic neurons back to the inhibitory cells that innervate them, suppressing inhibitory inputs for up to 10 seconds or more. DSI is particularly prominent in the hippocampus and cerebellum. Wilson and Nicoll (2001) suspected that a cannabinoid mechanism might be involved. They used slice preparations of rat hippocampus and induced DSI by depolarizing neurons with minute electrical currents via microelectrodes inserted into single cells. They found that DSI was completely blocked by the cannabinoid CB-1 receptor antagonists AM251 and rimonabant. They were also able to show by recording from pairs of nearby hippocampal neurons that depolarizing one of these neurons caused DSI to spread and affect adjacent neurons up to 20 μm away. The results suggested that the small lipid-soluble, freely diffusible endocannabinoids released from single neurons can act as retrograde synaptic signals that can affect axon terminals in a sphere of influence approximately 40 μm in diameter and persist for seconds rather than milliseconds. Although this is a minute volume of brain tissue, it would contain hundreds of individual neurons.

Further support for the conclusion that a cannabinoid-mediated mechanism underlies DSI came from Varma et al. (2001), who found that DSI was completely absent in hippocampal slices prepared from CB-1 receptor knockout mice. CB-1 receptors in the hippocampus are particularly abundant on the terminals of a subset of GABAergic basket cell interneurons that also contain the neuropeptide cholecystokinin (Katona et al., 1999), and these may well be the neurons involved in DSI. Retrograde signaling by cannabinoids is not restricted to the phenomenon of DSI; subsequent research has shown it to apply to DSE (a parallel process involving suppression of excitatory inputs) and to long-term depression (LTD), an inhibitory phenomenon lasting several minutes following brief stimulation of neurons in some parts of the brain (Alger, 2002). Endocannabinoids are also involved in long-term potentiation in hippocampal neurons, thought to be important in laying down of memories (Martinez and Derrick, 1996; Carlson et al., 2002). These findings suggest that endocannabinoids are involved in the rapid modulation of synaptic transmission in the central nervous system (CNS) by a novel retrograde signaling system causing local inhibitory effects on both excitatory and inhibitory neurotransmitter release that persist for tens of seconds or minutes. This may play a particularly important role in the control of the synchronized rhythmic firing patterns of neurons in hippocampus and elsewhere (Kano et al., 2009; Castillo et al., 2012). Externally administered THC or other cannabinoids cannot mimic the physiological effects of locally released endocannabinoids because the overall effect of administered cannabinoids is to cause a persistent inhibition of neurotransmitter release, not the transient effects seen with DSI and DSE.

Control of Energy Metabolism and Body Weight

Endocannabinoids in the CNS and in the periphery play a complex role in the control of obesity and energy metabolism (Carai et al., 2006; Matias and DiMarzo, 2007). It has long been known that cannabis users often experience sudden increases in appetite ("the munchies"), and this is probably due to an action of THC on CB-1 receptors in the hypothalamus. Hypothalamic levels of endocannabinoids are raised in hungry animals and in some animal models of obesity. Endocannabinoid actions in the hypothalamus involve complex interactions with hypothalamic peptides involved in appetite stimulation, or satiety (e.g., orexins, neuropeptide Y corticotropin-releasing factors, and melanocyte-stimulating hormone) (Lambert and Muccioli, 2007). The actions of cannabinoids in regulating body weight are not confined to the hypothalamic control of appetite, however. CB-1 receptors in fat tissues and in the liver seem to play an important role in regulating the synthesis of fats from foodstuffs (Osei-Hyiaman et al., 2005; Bellocchio et al., 2006 see Chapter 5, this volume). Endocannabinoid mechanisms in the pancreas are involved in the control of insulin secretion after feeding. In both pancreatic islets and adipose tissue, cannabinoids can promote the formation of new cells, adding to the size and responsiveness of these tissues. The cannabinoid influences on the balance between the satiety hormone leptin and the hunger-inducing hormone ghrelin are normally benign and are adaptive mechanisms capable of responding to alterations in the environment when food is scarce or more plentiful. In the modern world, in which there is an abundance of calorie-rich food, these mechanisms can malfunction and promote obesity and related metabolic disorders (Matias and Di Marzo, 2007). The importance of CB-1 receptor-mediated mechanisms is emphasized by the dramatic effects of the CB-1 receptor antagonist rimonabant in animals and in human subjects. By blocking these peripheral and central CB-1 receptor-mediated mechanisms, rimonabant reduced obesity and related metabolic disorders both in animal experiments and in human subjects (for details, see Chapter 5).

Regulation of Pain Sensitivity

Endocannabinoids reduce sensitivity to various types of pain (for reviews, see Walker and Hohmann, 2005; Zogopoulos et al., 2013). As outlined previously, cannabinoid receptors both in the CNS and in the periphery play a role in modifying pain responsivity. This situation is complicated in the case of the endocannabinoids because several of these compounds, notably anandamide and NADA, interact not only with cannabinoid receptors but also with another receptor that profoundly influences pain—the vanilloid receptor transient receptor potential vanilloid 1 (TRPV-1). This is found in the small-diameter sensory nerves that carry pain information into the CNS and in some parts of the brain. It is the target for capsaicin— the pungent principle of the Hungarian red pepper. Activation of the TRPV-1 protein by capsaicin causes intense pain by activating the pain-sensitive sensory nerves. Paradoxically, anandamide is also able to activate this mechanism, although not as potently or effectively as capsaicin (Spicaroval et al., 2014). The physiological significance of this dual action of endocannabinoids is unclear, but it may explain some of the apparently paradoxical effects of the compounds in animal models, in which anandamide can sometimes increase sensitivity to pain rather than cause analgesia. It is possible that the sensitivity of the CB-1 versus TRPV-1 targets to endocannabinoids may vary according to the pathological state of the animal.

One situation in which endocannabinoids are strongly activated appears to be stress, which is known to be capable of briefly dulling pain sensitivity (the wounded soldier or football player does not feel the pain immediately). In an animal model of stress-induced analgesia, much of the effect was blocked by rimonabant, and stress led to rapid accumulations of anandamide and 2-AG in an area of brainstem (periaqueductal gray) known to play a key role in regulating pain sensitivity (Morena et al., 2016).

Cardiovascular Control

As described in Chapter 3, CB-1 receptors are present in many blood vessels, and cannabinoids relax the smooth muscle of such vessels, causing drops in blood pressure. The endocannabinoids may achieve this in part by actions on CB-1 receptors, partly by activation of the vanilloid TRPV-1 protein and partly by triggering release of the vasodilator nitric oxide (Pacher et al., 2005). But although endocannabinoids can be made by the cells lining blood vessels, they do not seem to play an important role in the basal control of blood pressure. Treatment of animals or people with the CB-1 antagonist rimonabant does not affect blood pressure, and CB-1 receptor knockout mice have normal blood pressure. However, there is evidence that endocannabinoid mechanisms may become important in pathological states such as hypertension or in mediating the sudden drops in blood pressure that occur in conditions of shock—for example, after a sudden loss of blood (Pacher et al., 2005; O'Sullivan, 2015). Other results have suggested that anti-inflammatory effects of cannabinoids acting through CB-2 receptors could help protect against the development of atherosclerotic plaques in blood vessels—a key risk factor for heart disease (Steffens et al., 2005).

Other Functions

Many other roles have been described for endocannabinoids. For example, in the control of human reproduction, CB-1 receptors on the fertilized blastocyst and on the cells lining the uterus help control whether or not successful implantation of the blastocyst occurs. Excessive stimulation of these CB-1 receptors can block implantation or even precipitate early abortion (Friede et al., 2009). Local levels of anandamide are controlled by fluctuations in the activity of the degrading enzyme FAAH that are under the control of complex hormonal and other signals, including stress. It has been suggested that drugs that are able to enhance FAAH activity might be useful in treating human infertility (Warholak, 2016).

People take cannabis because of its pleasurable and rewarding effects (see Chapter 7). Studies of endocannabinoids and CB-1 receptor knockout mice are beginning to reveal the complex manner in which cannabinoids are involved in pleasure and reward pathways (Valverde et al., 2005). CB-1 knockout mice display heightened reactions to stress (Riebe and Wotjak, 2011), including increased fear and anxiety behavior when exposed to novel environments or other fearful stimuli. CB-1 receptor knockout mice are less able to forget painful unpleasant memories. When trained to associate a tone with an electric shock, they are slow to extinguish this memory when tone and shock are no longer paired (Mariscano et al., 2002). These mice also demonstrate the key role played by the CB-1 receptor in the development of dependence to THC: They do not display any of the withdrawal reactions seen in normal animals treated repeatedly with THC and then challenged with the antagonist rimonabant.

Endocannabinoids also play multiple roles in the gastrointestinal tract (Schicho and Storr, 2012; Lee et al., 2016); in the control of sleep–wake cycles (Prospero-Garcia et al., 2016); and potentially in the control of cancer—a number of in vitro studies have shown that cannabinoids can reduce cancer cell division and growth (Velasco et al., 2015; Ladin et al., 2016). However, none of these findings has yet been translated into human medicine.

Development of a New Endocannabinoid-Based Pharmacology

For reviews of the development of endocannabinoid-based pharmacology, see Piomelli (2003), Lambert and Fowler (2005), and Mackie (2006).

Novel Cannabinoid Receptor Agonists or Antagonists

The discovery of multiple cannabinoid receptors and the endocannabinoids as their natural ligands offers new opportunities for the development of selective agonists or antagonists. Synthetic compounds with selectivity as agonists or antagonists for CB-2 receptors are already known, and because the CB-2 receptor is only expressed at very low levels in the CNS, the problem of unwanted CNS side effects does not exist for such compounds. However, there has been relatively little research to identify the most suitable medical applications for CB-2 selective compounds. Far more promising has been the development of antagonists with selectivity for the CB-1 receptor. One such compound, rimonabant, is already approved as a novel treatment for metabolic disorders associated with obesity, and it is reviewed in more detail in Chapter 5.

Inhibitors of Endocannabinoid Inactivation

For reviews of inhibitors of endocannabinoid inactivation, see Di Marzo et al. (2005) and Mallet et al. (2016).

Knowledge of the biochemical mechanisms involved in the biosynthesis and inactivation of endocannabinoids has suggested a number of novel drug targets. The biosynthetic enzymes for 2-DAG, DAG lipase-α and DAG lipase-β, are potential targets, and inhibitors are available, notably tetrahydrolipostatin (half maximal inhibitory concentration (IC_{50}) = 60 nM), although there is little in vivo information. There has been far more interest in the discovery and development of compounds that inhibit the inactivation of endocannabinoids via the putative endocannabinoid transporter and the enzymes FAAH and DAG lipase. Such compounds would be expected to enhance endocannabinoid actions only in areas where there was already some activation of the cannabinoid system, making their effects far more selective and restricted than those of externally administered THC or related CB-1 receptor agonists. A particular focus of interest has been to develop inhibitors of the degradative enzyme FAAH. This enzyme is a member of a large family of related enzymes that degrade many different protein and amide substrates. As previously indicated, the enzyme has a complementary distribution in brain to that of the CB-1 receptor and is thought to play a key role in the inactivation of endocannabinoids after their transport into cells containing the enzyme. Strong evidence in support of this was provided by studying a FAAH knockout strain of mice, which exhibited increased levels of anandamide and proved to be supersensitive to anandamide (Cravatt et al., 2001). A number of laboratories have developed selective inhibitors of FAAH, which is not an easy task because the enzyme is quite similar to many other enzymes in the "amide hydrolase" family. Nevertheless, apparently selective FAAH inhibitors are now available (Figure 4.2), and their pharmacological properties are promising.

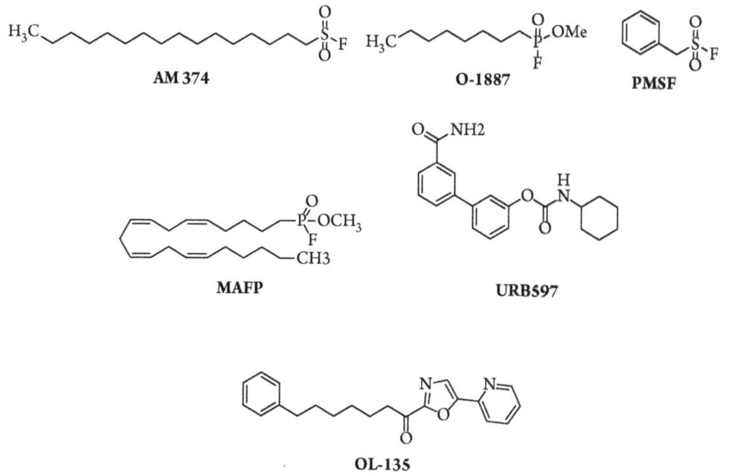

FIGURE 4.2 Selective FAAH inhibitors.

Source: From Pertwee (2006).

A notable molecule is LY 2183240, which is a very potent inhibitor of anandamide uptake (IC_{50} = 270 pM), but whether this reflects the potent inhibition of FAAH by this substance (IC_{50} = 12.4 nM) remains controversial. In whole animal experiments, these compounds do not themselves cause catalepsy, lowered body temperature, or increased food intake (Mackie, 2006; Piomelli et al., 2006). In experimental animals, in vivo administration of FAAH inhibitors enhances analgesic effects of cannabinoids, and on their own they have analgesic properties. However, clinical trials have not confirmed this—and the company involved has abandoned FAAH inhibitors. The profile of these compounds has also been clouded by the deaths and serious adverse effects in volunteers receiving one such compound in a Phase I trial (Enserink, 2016; Mallet et al., 2016). This disaster, and the failure of rimonabant as an anti-obesity drug (see Chapter 5), may put a damper on cannabinoid research by pharmaceutical companies for the time being.

Chapter 5

Medical Uses of Marijuana—Fact or Fantasy?

Historical

Cannabis has been used as a medicine for thousands of years (Lewin, 1931; Walton, 1938; Robinson, 1996). The Chinese compendium of herbal medicines, the *Pen-ts'ao*, first published around 2800 BC, recommended cannabis for the treatment of constipation, gout, malaria, rheumatism, and menstrual problems. Chinese herbal medicine texts continued to recommend cannabis preparations for many centuries. Among other things, its pain-relieving properties were exploited to relieve the pain of surgical operations.

Indian medicine has almost as long a history of using cannabis. The ancient medical text, the *Atharva Veda*, which dates from 2000–1400 BC, mentions "bhang" (the Indian term for marijuana), and further reference is made to this in the writings of Panini (approximately 300 BC) (Chopra and Chopra, 1957).

In the ancient Ayurvedic system of medicine, cannabis played an important role in Hindu materia medica, and it continues to be used by Ayurvedic practitioners today. In various medieval Ayurvedic texts, cannabis leaves and resin are recommended as decongestant, astringent, soothing, and capable of stimulating appetite and promoting digestion. Cannabis was also used to induce sleep and as an anesthetic for surgical operations. In addition, it was considered to have aphrodisiac properties and was recommended for this purpose.

In Arab medicine and in Muslim India, frequent mention is also made of "hashish" (cannabis resin) and "benj" (marijuana). They were used to treat gonorrhea, diarrhea, asthma, and as an appetite stimulant and analgesic. In Indian folk medicine, bhang and ganja (cannabis resin) were recommended as stimulants to improve staying power under conditions of severe exertion or fatigue. Poultices applied to wounds and sores were believed to promote healing or, when applied to areas of inflammation (e.g., piles), to act as an anodyne and sedative. Extracts of ganja were used to promote sleep and to treat painful neuralgias, migraine, and menstrual pain. Numerous concoctions containing extracts of cannabis together with various other herbal medicines continue to be used in rural Indian folk medicine today, with a variety of different medical indications, including dyspepsia, diarrhea, dysentery, fever, renal colic, dysmenorrhea, cough, and asthma. Cannabis-based tonics with aphrodisiac claims are also popular. The use of cannabis-based medicines has declined rapidly in India in recent years, however, as more reliable Western-style medicines have become more generally available.

Cannabis or *hemp* was also popular in folk medicine in medieval Europe and was mentioned as a healing plant in herbals such as those by William Turner, Mattioli and Dioscobas Taberaemontanus (Booth, 2003). One of the most famous herbals, written by Nicholas Culpepper (1616–1654), recommended for hemp seeds that "an emulsion of decoction of the seed . . . eases the colic and always the troublesome humours in the bowels and stays bleeding at the mouth, nose and other places" (Booth, 2003).

It was not until the middle of the 19th century, however, that cannabis-based medicines were taken up by mainstream Western medicine. This can be almost entirely attributed to the work of a young Irish doctor, William O'Shaughnessy, serving with the Bengal Medical Service of the East India Company (Booth, 2003). He had observed first-hand the many uses of cannabis in Indian medicine and had himself conducted a series of animal experiments to characterize its effects and establish what doses could be tolerated. His experiments confirmed that cannabis was remarkably safe. Despite many escalations of dose, it did not kill any of his experimental animals. O'Shaughnessy felt confident to go on to conduct studies in patients suffering from seizures, rheumatism, tetanus, and rabies. He found what appeared to be clear evidence that cannabis could relieve pain and act as a muscle relaxant and an anticonvulsant. The 30-year-old O'Shaughnessy reported his findings in a remarkable monograph, first published in Calcutta in 1839 and reprinted as a 40-page article in *Transactions of the Medical and Physical Society of Calcutta* in 1842 (O'Shaughnessy, 1842). His report rapidly attracted interest from clinicians throughout Europe. As a result of his careful studies, O'Shaughnessy felt able to recommend cannabis, particularly as an "anticonvulsive remedy of the greatest value." He brought back a quantity of cannabis to England in 1842, and Peter Squire in Oxford Street, London, was responsible for converting imported cannabis resin into a medicinal extract and distributing it to a large number of physicians under O'Shaughnessy's directions.

O'Shaughnessy was a remarkable Victorian genius (Moon, 1967). Before moving to India, he undertook a series of experimental inquiries in Newcastle-upon-Tyne into the composition of the blood in cholera and concluded correctly that there was dehydration and salt loss, and he advocated treatment designed to replace salt and hydrate patients. O'Shaughnessy never himself put these ideas to the test, but they were rapidly taken up by physicians and found to be effective. Cholera was a common and deadly infectious disease in 19th-century cities, which lacked modern sanitation systems. His ideas form the basis of "fluid replacement therapy," which to this day is the basis of treatment for the catastrophic loss of salts and water from the blood that is a key feature of cholera and other diseases that induce severe diarrhea. On moving to India in 1833, O'Shaughnessy began his studies of cannabis described previously, and in 1841 he published an important textbook of chemistry and was made professor of chemistry at the Medical College in Calcutta; 2 years later, at the remarkably young age of 34 years, he was elected a

Fellow of the Royal Society in London. He was subsequently instrumental in constructing thousands of miles of telegraph lines in India, which proved to be of critical importance for communication in this vast part of the British Empire. By the time that O'Shaughnessy retired to England in 1860, at the age of 51 years, there were 11,000 miles of telegraph lines in India and 150 offices in operation. Within 10 years, telegraph links to London would be established.

Following O'Shaughnessy's advocacy of cannabis and the availability of the medicinal extract, it became popular for a while in British medical circles. Many doctors began to experiment with cannabis as a new form of treatment, and reports appeared in medical journals describing its application in a variety of conditions, including menstrual cramps, asthma, childbirth, quinsy, cough, insomnia, migraine headaches, and even in the treatment of withdrawal from opium. Cannabis extract and tincture appeared in the *British Pharmacopoeia* and were available for more than 100 years

British Pharmaceutical Codex 1949:

EXTRACTUM CANNABIS

(Ext. Cannab.)

Extract of Cannabis:

Cannabis in coarse powder . . . . 1,000 g

Alcohol (90%) . . . . a sufficient quantity

Exhaust the cannabis by percolation with the alcohol and evaporate to the consistence of a soft extract. Store in well-closed containers, which prevent access of moisture.

Dose: 16 to 60 mg

TINCTURA CANNABIS

(Tinct. Cannab.)

Tincture of Cannabis:

Extract of Cannabis . . . . 50 g

Alcohol (90%) . . . . to 1,000 ml

Dissolve

Weight per ml at 20°, 0.842 g to 0.852 g

Alcohol content 83% to 87% v/v

Dose 0.3 to 1 ml

In Britain, the eminent Victorian physician Sir John Russell Reynolds (Reynolds 1890) recommended cannabis for sleeplessness, neuralgia, and dysmenorrhea (period pains). It was also experimented with as a means of strengthening uterine contractions in childbirth and in treating opium withdrawal, an increasing problem for Victorian medicine as the uncontrolled consumption of opium created problems of addiction. There was interest in the use of cannabis in the treatment of the insane, following reports by Dr. Jean Jacques Moreau in Paris of this possibility. But there was also concern that excessive use of cannabis could lead to insanity, a concern that persisted for many years—leading among other things to the Indian Hemp Drugs Commission's review of the use of cannabis in India at the end of the 19th century, and revived again recently with claims that the teenage use of cannabis may lead to subsequent mental illness (see Chapter 6).

Although Dr. Reynolds is said to have prescribed cannabis to Queen Victoria to treat her period pains, cannabis never really became popular in British medicine and was used only infrequently. Difficulties in obtaining supplies and the inconsistent results obtained with different preparations of the drug made it difficult to use. Because of the lack of any quality control to allow the preparation of standardized batches of the medicine, patients were likely to receive a dose that either had no effect or caused unwanted intoxication. Cannabis was never as reliable and widely used as opium— the mainstay of the Victorian medicine cabinet. Cannabis fell so far out of favor that it was the lack of any continuing medical use as much as any other factor that led to its removal from the *British Pharmacopoeia* by the middle of the 20th century.

In America, cannabis was already known even before O'Shaughnessy made it popular in Europe. It was first introduced into homeopathic medicine, as described in the *New Homeopathic Pharmacopoeia and Posology or the Preparation of Homeopathic*

Medicines (Jahr, 1842). Cannabis came to the notice of psychiatrists also, who experimented with its use in treating the mentally ill. By 1854, the US Dispensatory began to list cannabis among the nation's medicinals, and it gave the following remarkably accurate description of its properties:

> Medical Properties: Extract of hemp is a powerful narcotic, causing exhilaration, intoxication, delirious hallucinations, and, in its subsequent action drowsiness and stupor, with little effect upon the circulation. It is asserted also to act as a decided aphrodisiac, to increase the appetite, and occasionally to induce the cataleptic state. In morbid states of the system, it has been found to produce sleep, to allay spasm, to compose nervous inquietude, and to relieve pain. In these respects it resembles opium in its operation; but it differs from that narcotic in not diminishing the appetite, checking the secretions, or constipating the bowels. It is much less certain in its effects; but may sometimes be preferably employed, when opium is contraindicated by its nauseating or constipating effects, or its disposition to cause headache, and to check the bronchial secretion. The complaints to which it has been specially recommended are neuralgia, gout, tetanus, hydrophobia, epidemic cholera, convulsions, chorea, hysteria, mental depression, insanity, and uterine hemorrhage. Dr. Alexander Christison, of Edinburgh, has found it to have the property of hastening and increasing the contractions of the uterus in delivery, and has employed it with advantage for this purpose. It acts very quickly, and without anesthetic effect. It appears, however, to exert this influence only in a certain proportion of cases. (Wood and Bache, 1854)

Although cannabis continued to attract the interest of psychiatrists, it did not become widely popular with American doctors. During the Civil War, it was used to treat diarrhea and dysentery among the soldiers, but as a medicine cannabis had too many shortcomings. As British doctors had found, the potency of commercial preparations varied from pharmacist to pharmacist because there was no means of standardizing the preparations for their content of the active drug. In addition, the drug was not water soluble and so, unlike morphine, cannabis could not be given by injection. The hypodermic syringe was invented in the late 19th century and was immediately popular with doctors and patients for administering instant remedies. There is a certain mystique associated with the ritual of an injection—even today, many Japanese patients prefer their medicines to be administered in this way. Cannabis had to be given by mouth and took some time to take effect. The doctor might have to remain with his patient for more than an hour after giving the drug, in order to ensure not only that it was having the desired effect but also that the dosage had not been too high.

A succinct and perceptive summary of the rise and fall of cannabis in 19th-century medicine is given by Walton (1938):

> The popularity of the hemp drugs can be attributed partly to the fact that they were introduced before the synthetic hypnotics and analgesics. Chloral hydrate was not introduced until 1869 and was followed in the next 30 years by paraldehyde, sulfonal and the barbitals. Antipyrine and acetanilide, the first of their particular group of analgesics [aspirin-like drugs], were introduced about 1884 [aspirin, not until 1899]. For general sedative and analgesic purposes, the only drugs commonly used at this time were the morphine derivatives and their disadvantages were very well known. In fact, the most attractive feature of the hemp narcotics was probably the fact that they did not exhibit certain of the notorious disadvantages of the opiates. The hemp narcotics do not constipate at all, they more often increase rather than decrease appetite, they do not particularly depress the respiratory center even in large doses, they rarely or never cause pruritis or cutaneous eruption and, most importantly, the liability of developing addiction is very much less than with the opiates.

These features were responsible for the rapid rise in popularity of the drug. Several features can be recognised as contributing to the gradual decline of popularity. Cannabis does not usually produce analgesia or relax spastic conditions without producing cortical effects and, in fact, these cortical effects usually predominate. The actual degree of analgesia produced is much less than with the opiates. Most important, the effects are irregular due to marked variations in individual susceptibility and probably also to variable absorption of the gummy resin." (Walton, 1938, p. 152)

Pharmaceutical companies, nevertheless, tried to make use of cannabis as a medicine, and it was included in dozens of proprietary medicines that were available over the counter in the 19th century and the early years of the 20th century. These included the stomach remedy Chlorodyne (which also contained morphine; Squibb), Corn Collodium (Squibb); Dr. Brown's Sedative Tablets and One Day Cough Cure (Eli Lilly). The company Grimault and Sons marketed cannabis cigarettes as a remedy for asthma. By 1937, when cannabis was removed from medical use in the United States, approximately 28 different medicines contained it as an ingredient—many of them with no indication of its presence.

The Modern Revival of Interest in Cannabis-Based Medicines

For a review of the revival of interest in cannabis-based medicines, see Kendall and Yudowski (2017).

Introduction

During most of the 20th century, there was little interest in the use of cannabis in Western medicine, and such use has been legally prohibited since 1937 in the United States and since the 1970s in

Britain and most of Europe. In all of these countries, cannabis was classified as a Schedule I drug—that is, a dangerous addictive narcotic with no recognized medical uses. As cannabis became an increasingly popular recreational drug during the 1960s and 1970s, however, increasingly more people were exposed to it, and during the 1980s and 1990s there was an increasing interest in medical applications. Many normally law-abiding citizens in the developed world started to use cannabis illegally for therapy. The groups most commonly involved in such illegal self-medication were those suffering from chronic pain conditions unresponsive to other pain-relieving drugs. A survey of more than 2,000 self-selected users of medicinal cannabis in the United Kingdom showed that the most common indications were multiple sclerosis (MS), neuropathy, and other chronic pain states; arthritis; and depression(Ware et al., 2005). Similar results were obtained in a survey of the medical use of cannabis in the Netherlands, where it is legal (Gorter et al., 2005). It is worth noting that in medical applications, cannabis acts as a palliative rather than a cure.

The United States

California was the first state to approve the availability of medical marijuana on prescription in 2014; several other states have followed it.; state law permits the provision of medical marijuana in 31 states and the District of Columbia (Alaska, Arizona, Arkansas, California, Colorado, Connecticut, Delaware, Florida, Hawaii, Illinois, Maine, Maryland, Massachusetts, Michigan, Minnesota, Montana, Nevada, New Hampshire, New Jersey, New Mexico, New York, North Dakota, Ohio, Oregon, Pennsylvania, Rhode Island, Vermont, Washington, Washington DC, and West Virginia), and further states await the necessary changes in legislation (Indiana, Iowa, Kentucky, Missouri, Nebraska, North Carolina, Oklahoma, South Carolina, Tennessee, Texas, Utah, and Wisconsin) (ProCon, 2017).

Europe

A wide variety of different laws exist regarding medical marijuana, and availability also varies widely (European Monitoring Centre for Drugs and Drug Addiction, 2016). In several countries (including the United Kingdom), the medical product Sativex is approved but not available on the National Health Service. At one extreme, the Netherlands permits the legal availability of marijuana for both medical and recreational use; at the other extreme, in some Eastern European countries and in Russia, marijuana is illegal and its use punishable by draconian laws. In several European countries, marijuana remains illegal, but arrests or prosecutions for its use are rare. Cultivation of a limited number of cannabis plants for personal consumption is tolerated (Spain, the Netherlands, Czech Republic, and Belgium). Cultivation for personal use occurs in many other countries where it is illegal (e.g., the United Kingdom). Some individuals have sought to minimize the chance of prosecution by forming "cannabis social clubs" in which groups of patients club together to purchase and provide medical supplies of marijuana to members. Such clubs usually have written rules: marijuana for medical use only by members, and are non-profit organizations. Although these clubs are not formally recognized by any national governments, they are tacitly recognized in Spain, Belgium, the Netherlands, and Slovenia. In general, European countries appear to be inching grudgingly toward acceptance of legally available medical cannabis. The most tolerant countries are the Netherlands, Spain, and Czech Republic, followed by Italy, Germany, and Belgium. A review of the medicinal use of cannabis in Europe provides more detailed information, although the authors seek to distinguish medicinal cannabis (as an oral medicine) from smoked marijuana, stating (erroneously), "The problem with smoking cannabis is that it does not deliver the drug in a reproducible or predictable dosage, and does not provide any therapeutic benefit" (Bifulco and Pisanti, 2015).

Other Countries

Canada is an important country in which medical cannabis is legal and which has also recently legalized recreational cannabis. Some of the regulations and other arrangements put in place by the Canadian government are discussed in more detail later. Australia is another country in which cannabis has been widely available for some time. Throughout the world, countries in which patients have access to legal cannabis include Chile, Colombia, Jamaica, and Uruguay. Uruguay was the first country in the world to legalize the cultivation and sale of cannabis for medical and recreational use in 2014 (see Chapter 7).

Medicinal Cannabis

The evidence for an against medicinal cannabis has been the subject of a number of expert reviews in recent years. A report by the British Medical Association on the therapeutic uses of cannabis published in 1997 was followed by the UK House of Lords enquiry into cannabis (House of Lords, Select Committee on Science and Technology, 1998); the US Institute of Medicine's *Marijuana and Medicine* (1999); the British Royal College of Physicians' "Cannabis and Cannabis-Based Medicines" (2005); and reviews by Hill (2015), Whiting et al. (2015), and Barnes (2016). In the 1990s, the recognition that cannabis had genuine medical indications, supported by data from properly controlled clinical trials (Barnes, 2016), led to the official approval of cannabis in many countries in Europe and in some US states—starting with California in 1996. Since then, the approval of "cannabis pharmacies" now extends to 28 states and Washington, DC, with approval pending in another 12 in 2017 (ProCon, 2017). Each state has its own law regulating the amount of cannabis available, usually 2–8 ounces, and a limited number of cannabis plants for home cultivation. Patients must register, with written approval from their physician stating that they "may benefit from cannabis use." They should prove residency in the state in which they register and be older than age 18 years. On presenting their registration card at a local "cannabis pharmacy," they may obtain their supply of medical marijuana (ProCon, 2017).

The fact that cannabis has until very recently been an illegal substance has undoubtedly inhibited both preclinical and clinical research on its medical potential. Most clinical trials have been reported in the past 20 years, so rigorous scientific evidence either for its safety or for its effectiveness has not been available until recently, except in a few isolated instances. The introduction of new medicines for human use normally requires that they fulfill internationally agreed criteria for quality control, safety, and effectiveness, laid down by the various government regulatory agencies responsible for the approval of new medicines. Medical cannabis lies in a different realm. In most countries, there is little or no quality control of the product, and there is inadequate short-term or long-term evidence of toxicity. The greatest emphasis has been placed on the effectiveness of the medicine in treating a particular illness. This has to be established in controlled clinical trials (Box 5.1).

There are a number of variants on clinical trial design. For example, it is not always necessary to use separate groups of patients to assess test drug or placebo responses. In the so-called "crossover" trial design, the same patients receive placebo and test drug at different stages during the trial and are crossed over from one to the other after a "washout" period (to ensure removal of any active drug from the body). The test drug or placebo is given to different patients in random order so that the trial remains double blind.

These principles of clinical trial design, although they may appear to be simply common sense, are relatively new. It is only in the past 50 years that the concept of the "controlled clinical trial" has become generally accepted. It can be applied not only to the testing of new medicines but also to the effectiveness of any new medical procedure.

BOX 5.1 Controlled Clinical Trials

1. Comparing the test drug with an inactive placebo prepared in such a manner that it cannot be distinguished from the active medicine.
2. In a double-blind, randomized, placebo-controlled trial, neither the patient nor the doctor or nurse knows whether active drug or placebo is given to any particular patient.
3. Patients are randomly allocated to placebo and test drug groups to avoid any possible bias in the selection of those who are to receive the active drug. This information is held in coded form by a person not actively involved in the conduct of the trial and is not made available until the trial has ended.
4. The outcome of the trial should involve objective measurements wherever possible, using predetermined outcome measures or "end points."
5. The success or failure of the trial is measured by criteria established in a written trial protocol before the start of the trial.
6. Because individual patients will inevitably vary in their response to drug or placebo, the trial should include a sufficiently large number of subjects to provide statistically significant differences in outcome measures between the placebo and drug-treated groups. It is not uncommon for a clinical trial to involve hundreds or even thousands of subjects.

The reasons for insisting on these elaborate scientifically controlled trials was the growing realization that the expectations of both doctor and patient can influence the outcome of a clinical trial, even though neither may be consciously aware of this. The importance of the placebo effect means that this has to be built into the design of clinical trials. Not all human illnesses will show the same degree of susceptibility to the placebo effect; such treatment is most likely to affect the outcome of conditions in which there are strong psychological or psychosomatic components and less likely to influence the outcome of infectious diseases or cancer. Placebo effects are particularly prominent in the treatment of such psychiatric conditions as anxiety and depression, and they are often seen in illnesses in which the patient has failed to gain any benefit from existing conventional medicines. Such patients are often desperately seeking new treatments, which they want to work.

Fortunately, the past 20 years has seen a new era of controlled trials of cannabis-based medicines, yielding enough positive evidence to persuade government regulatory agencies in some advanced countries (Canada and Spain) to give formal approval for the use of these products.

Fully Licensed Cannabinoids

In the sound and fury of the debate about the medical use of cannabis, with strongly held positions on both sides of the argument, it is often forgotten that three cannabis-based medicines are already available on prescription to patients. These are the synthetic cannabinoids dronabinol (Marinol) and nabilone (Cesamet) and the herbal cannabis-derived product nabiximols (Sativex). The medical use of these products is backed up by a substantial body of scientific evidence from clinical trials, and the compounds have satisfied the strict requirements of the US Food and Drug Administration (FDA) and the corresponding Canadian and European agencies for approval as human medicines.

Dronabinol (Marinol and Syndros)

Dronabinol is the generic name given to synthetic delta-9-tetrahydrocannabinol (THC). It is marketed as the medical product known as Marinol. Drugs are given an official generic name, which is used when describing the compound in the scientific literature, and the company that markets the drug usually gives it a separate trade name. More than one company may market the same drug under different trade names, but each compound can only possess one generic name.

One of the problems in using dronabinol (THC) as a medicine is that the pure compound is a viscous pale-yellow resin, which is almost completely insoluble in water. This makes it impossible to prepare as a simple tablet, and it is not easily dissolved for administration as an intravenous injection. Marinol is prepared by dissolving dronabinol in a small quantity of harmless sesame oil in a soft gelatin capsule (containing 2.5, 5, or 10 mg dronabinol). These capsules are easily swallowed, and absorption is almost complete (90–95%), but because much of the active drug is metabolized during passage of the blood from the gut via the liver, only 10–20% of the administered dose reaches the general circulation. Effects begin after 30 minutes to 1 hour and reach a peak at 2–4 hours, with duration of action of 4–6 hours, although the appetite-stimulating effects of the drug may persist for up to 24 hours. Considerable quantities of the psychoactive metabolite 11-hydroxy-THC (see Chapter 2) are formed in the liver, and this metabolite is present in blood at approximately the same level as the parent drug, with a similar duration of action. Syndros is a liquid formulation of THC, administered by an oral mucosal spray. Because of the dense vasculature of the oral cavity, Syndros is absorbed more quickly than Marinol, and the dose can more easily be adjusted, although the two preparations are bioequivalent in terms of overall absorption.

Two medical indications have been formally approved for Marinol and Syndros, backed by clinical trial data. The first of these is their use to counteract the nausea and vomiting frequently associated with cancer chemotherapy. The second is as an appetite stimulant to counteract the AIDS wasting syndrome, as described later (for review, see Plasse et al., 1991). Approximately 80% of prescriptions are for HIV/AIDS patients, 10% for cancer chemotherapy, and 10% for a range of other purposes. Syndros was approved by FDA in

2017. Annual sales of Marinol peaked at $150 million in 2011 and then declined after the patent expired and a generic form became available; the market for Syndros is limited by the perceived danger of abuse.

The possibility that medical supplies of Marinol might be diverted to illicit use has been a concern, but there is very little evidence that this has happened. Marinol has little value as a street drug. The onset of action is slow and gradual, and its effects are unappealing to regular marijuana smokers; it has a very low abuse potential. Because of the low dependence liability, the US Drug Enforcement Agency reclassified Marinol to the less restrictive Schedule III in 1998, although pure THC remains a Schedule I substance. Syndros was considered by the FDA to carry a somewhat higher risk of abuse and was given a Schedule II label.

Nabilone (Cesamet)

During the 1970s, a number of pharmaceutical companies carried out research on synthetic analogs of THC to determine whether it might be possible to dissociate the desired medical effects from the psychotropic actions. On the whole, this quest proved disappointing, and in retrospect this may have been inevitable because we now believe that both the desired effects and the intoxicant actions of THC result from activation of the same CB-1 receptors in the central nervous system (CNS). Only one company persisted with this research long enough to produce a marketed product—nabilone (Cesamet). Nabilone is a potent analog of THC; it is a stable crystalline solid, and for human use, the drug is prepared in solid form in capsules containing 1 mg of nabilone that are taken by mouth, and the dose is usually 1 or 2 mg twice a day. Its effectiveness in the treatment of nausea and vomiting in patients undergoing cancer chemotherapy was demonstrated by a rigorous series of controlled clinical trials (for review, see Lemberger, 1985). Cesamet was approved by the FDA in 1985, but it was abandoned by Eli Lilly in 1989 for commercial reasons. Cesamet was purchased by Valeant Pharmaceutical in 2004, and after changes to the drug's labeling, it was marketed in Canada, and subsequently in the United States, in partnership with Par Pharmaceutical, after FDA approval in 2006. The product carries a Schedule II label, suggesting the possibility of abuse.

Nabiximols (Sativex)

Nabiximols (Sativex) is an extract of cannabis developed as a pharmaceutical product. It is extracted from cannabis plants grown under optimal conditions of light and temperature and free of insect or microorganism infections (Potter, 2014). The herbal material is extracted with liquid carbon dioxide, and the extract is formulated in a liquid containing a small amount of alcohol with propylene dioxide and peppermint oil. The principal active cannabinoid components are the cannabinoids THC) and cannabidiol (CBD). The liquid is provided in metered dose vial; each spray delivers 100 µl containing a dose of 2.7 mg THC and 2.5 mg CBD (https://www.gwpharm.com/products-pipeline/Sativex). Because the conditions of production and formulation are standardized, the cannabinoid content is very consistent. Sativex fulfills the conditions of standardization, formulation, and safety required of a novel pharmaceutical product for human use. It was approved as a botanical drug in the United Kingdom in 2010, as a mouth spray to alleviate spasticity associated with MS. Canada also formally approved nabiximols for the treatment of spasticity in MS and included approval for the treatment of pain in MS patients (Keating, 2017). Nabiximols is available in a number of countries as an unlicensed medicine, which enables doctors to prescribe the product to people who they consider may benefit. The product has been exported from the United Kingdom to a total of 28 countries, including Canada. In the United States, Sativex is not yet an approved medicine, although it is in advanced stages of clinical trials and FDA approval for cancer-related pain. In the United Kingdom, the Medicines and Healthcare Products

Regulatory Agency (MHRA) issued a marketing authorization for Sativex in 2010, but the National Institute for Clinical Excellence (NICE) did not recommend Sativex for reimbursement by the National Health Service (NHS) because it was considered to be too expensive for a modest clinical benefit (NICE, 2014). Many people with MS in the United Kingdom cannot receive Sativex due to local NHS resistance to its funding, as a form of postcode lottery. Regulatory agencies have proved reluctant to grant full approval to a medicine that is essentially a liquid form of cannabis, although instances of misuse of the product are rare.

Medical Indications for Cannabis

In the United States, each state that has allowed the use of medical marijuana specifies a list of medical indications, but these are fairly broad. Indications that are approved by 80% of medical marijuana states are "Alzheimer's disease, amyotrophic lateral sclerosis, cachexia/wasting syndrome, cancer, Crohn's disease, epilepsy and seizures, glaucoma, hepatitis C virus, human immunodeficiency syndrome, multiple sclerosis muscle spasticity, severe and chronic pain, and severe nausea" (Belendiuk et al., 2015, p. 10).

Unfortunately, scientific findings indicate that there is insufficient evidence to support the recommendation of medical marijuana at this time for the majority of the range of indications listed by US states as suitable for medical marijuana (Belendiuk et al., 2015; Hill, 2015; Whiting et al., 2015; Barnes, 2016; National Academies of Sciences, Engineering, and Medicine, 2017). The National Academies detailed report concluded,

> Despite the extensive changes in policy at the state level and the rapid rise in the use of cannabis both for medical purposes and for recreational use, conclusive evidence regarding the short- and long-term health effects (harms and benefits) of cannabis use remains elusive.

Multiple Sclerosis

Multiple sclerosis is the most widespread disabling neurological condition of young adults throughout the world, with an estimated 165,000 living with the condition in the United Kingdom, more than 400,000 living with it in the United States, and a worldwide total of approximately 2.3 million. Multiple sclerosis is a progressive, degenerative disease in which the brain and the spinal cord nerves are damaged by the gradual destruction of myelin, the protective, insulating layer of fatty tissue that normally coats nerve fibers. The precise cause of the disease is not known, but it is thought to represent an "autoimmune" condition, in which the body's immune system becomes inappropriately sensitized to some component of myelin—leading to its attack and progressive damage by the immune system. The disease usually progresses in stages, with periods of remission between, but it is ultimately life threatening. The symptoms are very variable, depending on which particular nerves or regions of the CNS are damaged, but it often manifests itself with symptoms of muscle spasticity, pain, and bladder and bowel dysfunction.

The treatment of MS has improved radically in recent years. With the introduction of new medicines that slow the rate of progression of the disease, beta-interferon, the antibody natalizumab, and more recently introduced disease-modifying monoclonal antibodies, more than a dozen such medicines have been approved by the FDA and are available (Jeffrey, 2013); however, none of these represents a complete cure (National Multiple Sclerosis Society, 2017). There are also several medicines available to treat the symptoms of MS, but none are wholly effective. The drugs baclofen and diazepam (Valium) help relax muscle spasms by activating receptors for the inhibitory chemical messenger GABA in brain and spinal cord, thus counteracting overactivity in the flow of excitatory signals to muscles. Chronic pain in MS sufferers is often difficult to treat, but drugs used in the treatment of epilepsy (carbamazepine and phenytoin) or depression (amitriptyline) and even opiates are sometimes used.

Multiple sclerosis represents a promising target for cannabis-based medicines (Butterfield, 2016). Anecdotal reports from self-medicating patients suggested that cannabis could not only relieve the muscle spasms and the pain that they cause but also, in some patients, improve bladder control. In 1998, the British Multiple Sclerosis Society reported the results of a survey of their 35,000 members to the House of Lords' Select Committee on Science and Technology indicating that a majority of patients who used (illegal) cannabis experienced some benefit. The use of cannabis in the treatment of various types of painful muscle spasms has a sound scientific rationale. Cannabinoid CB-1 receptors are found in particularly high density in those regions of the brain that are involved in the control of muscle function—the basal ganglia and the cerebellum. The receptors are densely located on output neurons in the outflow relay stations of the basal ganglia (substantia nigra and globus pallidus), where they are well placed to affect the control of movements. Activation of the cannabinoid receptors is known to suppress movements and can lead to a condition of catalepsy, in which the person or animal may remain conscious but immobile for considerable periods. It is not surprising, therefore, that cannabinoid drugs possess anti-spastic properties. In an animal model of MS in mice (allergic encephalomyelitis) in which the animal's immune system is sensitized to a component of its own myelin and there is progressive nervous system damage accompanied by muscle tremor, the symptoms could be suppressed by treatment with THC (Wirguin et al., 1999).

The successful treatment of spasticity in MS patients with nabiximols (Sativex) was a landmark in the treatment of a medical condition with cannabis. The efficacy of Sativex has been demonstrated in several large-scale controlled clinical trials (Barnes, 2006; Vermersch, 2011; Syed et al., 2014; Barnes, 2016). Sativex was formally approved for this indication and associated pain in Canada in 2005, and it was approved for spasticity in the United Kingdom in 2010 and also in several other European countries. In the United

Kingdom, many people with MS cannot receive Sativex due to local NHS resistance to its funding, as a form of postcode lottery. Some regulatory agencies have proved reluctant to grant full approval to a medicine (Sativex) that is essentially a liquid form of cannabis, although instances of misuse of the product are rare. Vermersch (2011) noted that "randomized controlled clinical trials of Sativex as add-on therapy provide conclusive evidence of its efficacy in the treatment of more than 1500 patients with multiple sclerosis (MS)-related resistance spasticity. In addition to those with MS, marijuana is also used illegally by other groups of patients who suffer from disabling illnesses that are accompanied by painful muscle spasms, including cerebral palsy, torticollis, various dystonias, and spinal injury. Survey data from patients suffering from spinal injuries indicated that more than 90% reported marijuana helped improve symptoms of muscle spasms of arms or legs and improved urinary control and function, but there are no scientific data to support the efficacy of cannabinoids in any of these conditions (Vermersch, 2011).

Pain

As reviewed in Chapter 3, there is a body of evidence from experiments in animals that indicates that activation of the cannabinoid system in the CNS, among other things, reduces the sensitivity to pain. Clinical pain comes in many varieties, from the severe but usually short-lived pain that follows injury or surgical operation to the chronic and often disabling pain that often accompanies such illnesses as rheumatism and arthritis or cancer.

Many different analgesic (pain-relieving) medicines are available, from aspirin and the many aspirin-like anti-inflammatory drugs that act on peripheral inflamed tissues to morphine, codeine, and other opiates that act directly on the CNS. None of them are completely satisfactory. Use of aspirin-like drugs carries with it the danger of irritation and ulceration of the stomach, which can lead to dangerous internal bleeding. Several thousand people die each year because of these side effects. Morphine and other opiates often cause severe constipation and at high doses can depress respiration and cause death. The repeated use of opiates can lead to the development of tolerance so that patients become increasingly less sensitive to the drugs and require increasing doses; some may become dependent on the opiate. As with cannabis, the psychotropic effects of opiates are disturbing rather than pleasurable to most patients. Nowadays, many patients are provided skin patches that deliver fentanyl or other potent morphine-like opiates transdermally in a continuous manner.

Some of the most distressing forms of clinical pain are those that arise from damage to nerves or to the spinal cord or brain. This can arise from many different causes: as a consequence of accidental or surgical injury to nerves; in patients with diabetes or AIDS, which often lead to damage in peripheral nerves; in MS as described previously; as a result of treatment with powerful cancer chemotherapy drugs that can damage nerves; and in some forms of cancer in which the tumor presses on or damages nerve fibers. These so-called "neuropathic" pain syndromes are often long-lasting and severe—and they are very difficult to treat because even the most powerful analgesic drugs, the opiates, are generally ineffective. In some cases, patients respond to treatment with anti-epilepsy drugs such as carbamazepine, phenytoin, or gabapentin or to drugs used more commonly in the treatment of depression, such as amitriptyline. For many sufferers, however, neuropathic pain remains untreatable.

An encouraging feature of the results on animal models is that THC has been reported to be effective as an analgesic in models of neuropathic pain. For example, in rats in which the sciatic nerve (which innervates the hind limb) is damaged surgically, the partially denervated limb becomes sensitized to pain (allodynia). Morphine and related opiates have previously been shown to be ineffective in this and other animal models of neuropathic pain. The allodynia lowering induced by the local activation of endocannabinoids suggests that they may play an important role in modulating neuropathic pain (Luongo et al., 2017).

The historical literature on the medical uses of cannabis has also long stressed its value in the treatment of a variety of painful conditions, but there have been few large-scale controlled scientific studies (Campbell et al., 2001; Barnes, 2016). There is no objective measure of pain. Clinical trials must therefore always rely on the patient's own reports. Commonly used subjective measures include numerical and visual analog scales, often with an 11-point scale from "no pain" to "worst possible pain," which yield daily pain scores for each patient. The subjective nature of the measurements makes clinical studies liable to a considerable placebo effect (Figure 5.1).

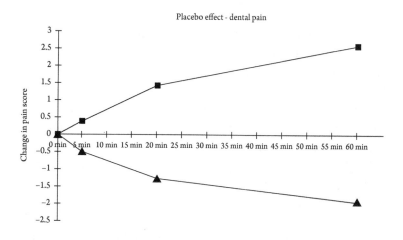

FIGURE 5.1 Placebo effect. Subjects received an injection of either morphine or saline (placebo) in a blinded manner 2 hours after dental surgery and were asked to rate their pain scores on an arbitrary scale for the subsequent hour. Data are shown only for the group receiving placebo. Of 107 patients in this group, 42 (39%) were rated as placebo responders. Whereas non-responders experienced an increasing level of pain (solid squares), the placebo responders either reported some degree of pain relief (solid triangles) or their pain remained unchanged (data not shown).

Source: Redrawn from Levine et al. (1979).

Relatively small-scale controlled clinical trials demonstrated a positive effect of Sativex in patients with chronic neuropathic pain (Serpell et al., 2005; Finnerup, 2014). Larger clinical trials have been completed with Sativex in the treatment of neuropathic pain (Numikko et al., 2007; Russo, 2008; Langford et al., 2013). Although Sativex led to a reduction in pain scores, the results were sometimes not statistically significant because of a large and variable placebo response. Clinical trials in pain associated with advanced cancer have shown a significant effect of Sativex in reducing pain, potentially leading to FDA approval for this indication in the United States (Johnson et al., 2010; Portenoy et al., 2012). Sativex is already formally approved in Canada for the treatment of neuropathic pain associated with MS (2005) and for cancer-associated pain (2007). These studies have provided scientific evidence to support the historical claims for the effectiveness of cannabis in pain relief. Formal approval of Sativex is likely to occur eventually in response to an ever-increasing body of positive controlled trial data (Barnes, 2006; Russo, 2008).

The available data suggest that cannabis-based medicines can benefit patients with certain difficult-to-treat pain conditions. Although their efficacy is variable and somewhat limited (1 or 2 points on a 10-point scale), patients often report the effects as clinically meaningful. Not all patients benefit (as witness the relatively high dropout rates due to lack of efficacy). Intoxication and other adverse side effects appear modest and well tolerated, particularly when patients can adjust their own optimum dose. Barnes (2016) reviewed the effects of Sativex (nabiximols) and other cannabinoids in treating pain and concluded that

> nabilone, dronabinol, nabiximols and smoked marijuana have all been shown to be efficacious to varying extents in a variety of pain settings in good quality studies. We conclude that there is good evidence for efficacy of cannabis for pain relief in various formulations and in a number of settings.

An important role for CB-2 receptors in pain control is suggested by various animal tests of CB-2 selective agonists (e.g., JWH-133 and GW-40593) showing their efficacy in various animal models of inflammatory pain (Malan et al., 2003). For example, Burston et al. (2013) reported that JWH-133 prevented the development of knee joint osteoarthritis-related pain behavior in an animal model. A study of the effects of the chronic activation of cannabinoid receptors found that the ability of THC to reverse the painful allodynia caused by the cancer chemotherapy agent paclitaxel in mice was mimicked by the selective CB-2 receptor agonist AM1710. The effects of AM1710 were absent in CB-2 receptor knockout mice (Deng et al., 2015). Evidence for CB-2 receptor anti-inflammatory actions in the rat brainstem reinforces the importance of CB-2 receptors in inflammatory pain (Li et al., 2016), despite the failure of several clinical trials of CB-2 selective ligands in the treatment of pain (Dhopeshwarkar and Mackie, 2014). The attractiveness of CB-2 selective drugs is due to their lack of psychoactive effects. Although this field of research holds considerable promise, the only clinical data so far are from small-scale trials.

Nausea and Vomiting Associated with Cancer Chemotherapy

One of the most distressing symptoms in medicine is the prolonged nausea and vomiting which regularly accompanies treatment with many anti-cancer agents. This can be so severe that patients come to dread their treatment; some find the side-effects of the drugs worse than the disease they are designed to treat; others find the symptoms so intolerable that they decline further therapy despite the presence of malignant disease.

— BRITISH MEDICAL ASSOCIATION *(1997, p. 21)*

Ironically, this condition for which there was the earliest scientific evidence for beneficial effects of cannabis-based medicines is now no longer viewed as an area of pressing medical need because new and even more powerful anti-sickness drugs have become available. Currently, 48 prescription medicines are available for the indication of vomiting (Drugs.com, 2016). When the cannabinoids dronabinol and nabilone were first being tested in the 1970s and early 1980s, however, matters were different.

Among the most effective anti-cancer drugs are the platinum-containing compound cisplatin, the plant product taxol, and numerous modern versions; unfortunately, they are also very powerful in causing nausea and vomiting. Cancer patients receiving these drugs almost invariably experience nausea and vomiting, with an average of six bouts of vomiting during the first 24 hours, unless they are protected by anti-emetic medicines. The initial reaction is followed by a delayed phase of nausea and vomiting during the next few days. The results of properly controlled clinical trials conducted in the 1970s and 1980s indicated that the two cannabinoid drugs dronabinol and nabilone appeared to offer a potentially important advance over the relatively ineffective anti-sickness medicines available in the early 1980s (Abrahamov et al, 1995; for review, see Tramer et al., 2001). The most widely used drugs then were chlorpromazine, prochlorperazine, haloperidol, metoclopramide, and domperidone—all of which act as antagonists of the chemical messenger dopamine. Various clinical trials in which dronabinol was compared with placebo or with another anti-sickness agent, prochlorperazine, reported the efficacy of dronabinol (for review, see Mücke et al., 2016).

The manufacturer of nabilone, Eli Lilly, conducted approximately 20 separate clinical trials involving more than 500 patients, many with a double-blind crossover design to allow the direct comparison of nabilone with prochlorperazine or other anti-emetic medicines in the same patients. Nabilone proved to be as effective as prochlorperazine, or more so, and it successfully treated the symptoms of nausea and vomiting in 50–70% of patients. Central nervous system side effects of drowsiness, light-headedness, and dizziness were seen in more than half of the patients, but these were not considered serious, and only a small proportion of patients (approximately 15%) experienced a "high" (Lemberger, 1985). The FDA granted formal approval of nabilone as an anti-emetic without causing intoxication, but the US Drug Enforcement Agency concluded that nabilone was still too much like cannabis, and it gave nabilone a restrictive Schedule II classification—that is, it was considered to be a potentially dangerous drug of addiction, although it did have some medical usefulness. The Schedule II classification was disappointing to Eli Lilly because it placed onerous requirements on the company and any physicians using the compound to keep it securely and to record its every movement. The company lost interest in the compound and in further research.

Epilepsy

Epilepsy is a major medical indication of unmet need; treatment-resistant epilepsy affects 30% of epileptic patient's and is associated with severe morbidity and increased mortality. Cannabis has been used to treat epilepsy for thousands of year, and there is a wealth of preclinical evidence for its efficacy (Devinsky et al., 2014), but until recently, the only clinical evidence for efficacy came from small trials or case reports (Devinsky et al., 2014; O'Connell et al., 2017). A plant-derived product of one of the naturally occurring cannabinoids, cannabidiol (Epidiolex), is in advanced development for treatment of rare forms of childhood epilepsy and may have more general application as an anti-epileptic drug. In an open-label trial investigating the effects of CBD (Epidiolex, 2–50 mg/kg) on 214 young patients with treatment-resistant epilepsy, CBD reduced seizure frequency and demonstrated an acceptable safety profile (Devinsky et al., 2017). Two large Phase III clinical studies of Epidiolux confirmed it to be active in treating childhood epilepsy (Lennox–Gastaut syndrome and Davet syndrome) (Devinsky et al., 2017). Cannabidiol is active against various animal models of epilepsy (Scuderi et al., 2009). Epidiolux is a cannabidiol product derived from plant strains that contain cannabidiol without THC (see   https://www.gwpharm.com/epilepsy-patients-caregivers/ patients). The positive effects reported in these trials will probably gain FDA approval for this product in childhood epilepsies, and the FDA has already given partial approval. Because treatment-resistant epilepsy is common, and cannabidiol lacks the limiting psychoactive features of THC, this is an area of high commercial interest.

Appetite Stimulation

Loss of appetite and a progressive involuntary weight loss of approximately 10% of body weight are seen in AIDS wasting syndrome, a characteristic feature of the disease. The onset of bouts of wasting syndrome, which last for a month or more, is one of the defining events in the transition from HIV-positive to AIDS. The wasting is accompanied by chronic diarrhea, weakness, and fever. The advent of the newer and more powerful treatments for HIV/AIDS may make the wasting syndrome less common in the future, but it remains a distressing feature of the disease. Although the precise physiological mechanisms underlying the wasting syndrome are not well understood, the loss of weight seems to be due primarily to reduced energy intake. After a series of small-scale clinical trials gave promising results, a larger placebo-controlled clinical trial was conducted in 139 such patients (Beal et al., 1995). Dronabinol was formally approved by the FDA as an appetite stimulant to treat the loss of appetite and weight associated with AIDS. Some clinical studies have indicated that dronabinol causes a significant stimulation of appetite in cancer patients, who also commonly suffer loss of appetite and an accompanying body weight loss. As in the treatment of nausea and vomiting, the principal adverse side effect limiting the use of dronabinol as an appetite stimulant has been the accompanying CNS side effects. Whether there is a broader utility for cannabinoids in stimulating appetite in cancer patients or in other medical indications is unclear. Barnes (2016) concluded,

In summary, we consider there is moderate evidence for improvement in appetite and for weight gain in AIDS patients. There is much less satisfactory evidence for similar improvements in cancer patients but nevertheless there is a little patchy evidence of efficacy but more and larger studies are required. There is no convincing evidence of efficacy for appetite stimulation in anorexia nervosa but adequate studies have not been done.

Other Potential Medical Indications

Mood Disorders

Cannabis has been advocated as a treatment for depressive illness, anxiety, and sleep disorders. One of the first recommended uses of cannabis in Western medicine was for the treatment of depressive illness and melancholia, and before the discovery of modern antidepressant drugs, cannabis continued to be used in this way during the first half of the 20th century. However, the few clinical trials that have been conducted with THC or nabilone in the treatment of depressive illness or anxiety have had mixed results. Although some patients reported improvements, others found the psychic effects of the cannabinoids unpleasant and frightening. Rather than relieving anxiety, the acute effect can be to provoke anxiety and panic in some subjects—particularly those who have had no previous exposure to cannabis. Witkin et al. (2005) put forward the counterargument that the activation of CB-1 receptors on nerve terminals in the brain suppresses the release of serotonin and other monoamines thought to be important in preventing depressive illness and also suggested that use of antagonists of CB-1 receptors might be a better approach to the treatment of depressive illness.

Acute stress increases anxiety-like behavior in rats via an endocannabinoid-dependent mechanism, suggesting a role for endocannabinoid mechanisms in stress-induced anxiety (Di et al., 2016). The efficacy of cannabis in treating post-traumatic stress disorder has been claimed (Butterfield, 2017), and some US states include this among the medical indications for treatment with cannabis; however, there is currently no scientific evidence to support these claims (Barnes, 2016). For a systematic review of the evidence for cannabis as an anxiolytic, see Crippa et al. (2009), and for preclinical data, see Bruijnzeel et al. (2016).

Sleep

Despite the widespread use of marijuana to promote sleep, there are few controlled trials (Babson et al., 2017). In sleep laboratory studies, orally administered THC at doses of 10–30 mg has been shown to cause increases in "deep" slow wave sleep, but at the same time—as with other hypnotic drugs—there is a decrease in "dreaming" or rapid eye movement (REM) sleep. After repeated treatment with large doses of THC, there was evidence of some degree of hangover during the morning after treatment and a rebound in the amount of REM sleep. THC thus does not appear to offer any advantages over existing sleeping pills, and it has the disadvantage of causing intoxication prior to sleep. Although Sativex has been found to reduce sleep disturbance in patients suffering from MS or chronic pain (discussed previously), this is probably because it relieves the underlying symptoms, notably chronic pain, leading to sleep disturbance rather than due to any direct effect on sleep mechanisms (Russo et al., 2007).

Neurodegenerative Diseases

Overall, although there is a theoretical basis for cannabinoids to provide neuroprotection, there is not yet convincing evidence of efficacy in the context of traumatic brain injury, Parkinson's disease, or Huntington's chorea (Barnes, 2016). Particular interest has focused on cannabidiol in the treatment of neurodegenerative disorders because of its anti-inflammatory and antioxidant properties (Fernández-Ruiz et al., 2013).

Migraine Headache

Most studies have been too small. A moderately sized clinical trial of migraine headache gave mixed results: "In summary, it is surprising that despite a long history of use in headache and migraine there are no good quality randomised clinical trials and thus no conclusion can be drawn" (Barnes, 2016).

Inflammatory Bowel Disease

Marijuana and other cannabinoids are frequently used to treat diarrhea, abdominal pain, and Crohn's disease (an inflammatory disease of the colon), but what is the evidence? An expert review of the clinical trials in this area suggested that there was "a lack of clinical studies to prove efficacy, tolerability and safety of cannabinoid-based medication for IBD [inflammatory bowel disease] patients" (Hasenoehrl et al., 2017).

Is There Any Role for Smoked Marijuana as a Medicine?

Given the well-documented adverse effects of smoked marijuana on the lungs (see Chapter 6), is there any place at all for smoked marijuana in medicine? Apart from the potential respiratory hazards, the idea of a smoked herbal remedy goes against the grain of much of our thinking in scientifically based medicine. As the American Medical Association (1997) stated,

> The concept of burning and inhaling the combustion products of a dried plant product containing dozens of toxic and carcinogenic chemicals as a therapeutic agent represents a significant departure from the standard drug approval process. According to this viewpoint, legitimate therapeutic agents are comprised of a purified substance(s) that can be manufactured and tested in a reproducible manner.

Nevertheless, a majority of state sanctioned medical cannabis users in the United States use smoking as the preferred route of administration, although the use of vaping is increasing. In Canada, use of a vaporizer is preferred by a majority of patients (53%), followed closely by smoking (47%) (Shiplo et al., 2016). There is little doubt that smoking provides a more reliable means of delivering THC than taking THC or cannabis extracts by mouth. Because of the variable and delayed absorption of orally administered THC, the patient is always exposed to the possibility of either under- or overdosing. Smoking, on the other hand, with some practice, permits the rapid delivery of what the individual patient judges to be the correct therapeutic dose. The oromucosal spray nabiximols (sometimes referred to as "liquid cannabis") is a compromise between smoking and the oral route, but absorption is not particularly rapid. Some advances have been made in the development of aerosol formulations of THC and vaporizers as alternative delivery systems for cannabis (see Chapter 2) that have been approved for some medical applications.

A few clinical trials have attempted to assess the effectiveness of smoked marijuana; for example, in controlling the symptoms of nausea and vomiting in patients undergoing cancer chemotherapy. Some studies have used "placebo" marijuana cigarettes, using herbal cannabis from which THC had been extracted with alcohol beforehand. Experienced marijuana users, however, have little difficulty in distinguishing the THC-containing smoked material from the placebo, making it difficult to undertake a properly blinded trial. Partly because of such difficulties, very few controlled clinical trials have ever been described (see the National Institutes of Health's 1997 "Report on the Medical Uses of Marijuana" and the American Medical Association's "Report on Medical Marijuana") A critical meta-analysis of individual patient data from five randomized trials of smoked marijuana in neuropathic pain was sponsored by the American Pain Society in 2015. The report concluded that inhaled marijuana may provide short-term relief for one of five or one of six patients with neuropathic pain (Andreae et al., 2015).

Cannabinoid Antagonist for the Treatment of Obesity

The involvement of cannabinoid mechanisms in the control of food intake and body weight has been referred to previously: Cannabinoid agonists stimulate appetite. The discovery that the antagonist rimonabant was capable of blocking the cannabinoid mechanisms both in the brain and in the periphery prompted clinical trials of this drug in the treatment of obesity, and the positive results of these trials led to its approval in Europe in 2006 as an important new medicine for combating the ever-increasing problem of obesity in Western societies. The history of the rapid development of rimonabant since the first scientific publication in 1994 is a good example of how preclinical data can suggest and guide a clinical development program (reviewed by Carai et al., 2006). A number of studies showed that rimonabant caused a marked reduction in daily food intake when given to normal or obese rats and mice given unlimited access to normal or high-fat diets. It was particularly effective in reducing the intake of palatable foods—normally consumed avidly even by satiated rats (e.g., condensed milk and chocolate-flavored drinks). This was accompanied by significant reductions in body weight. However, the effects of rimonabant on food intake diminished with repeated dosing and were no longer seen after the first week. Despite this, the drug continued to cause reductions in body weight, even though food intake had recovered to near normal levels. This could be explained by the finding of increased energy expenditure in the treated animals. A key target seems to be the fat tissue, whose cells carry CB-1 receptors. Blockade of these receptors led to increased synthesis and release into the circulation of the hormone adiponectin, which plays an important role in energy balance. Adiponectin stimulates the metabolism of fatty acids (otherwise deposited as fat), decreases plasma glucose levels, and decreases body weight. CB-1 receptors in the liver may also be involved because activation of these receptors stimulates fatty acid synthesis and promotes diet-induced obesity (Osei-Hyiamin et al., 2005). Rimonabant treatment of animals made obese by overfeeding showed decreased amounts of fat tissue; increased energy expenditure and fat breakdown; normalized plasma levels of glucose; reduced plasma levels of triglycerides (fat); decreased "bad" low-density lipoprotein cholesterol; and decreased the otherwise abnormally high amounts of another key hormone, leptin, in plasma. Leptin is made by fat cells and secreted into the circulation. In the brain, it acts on the hypothalamus to cause a reduction in food intake, as part of the complex mechanisms whereby the brain helps control food intake and body weight (Morton et al., 2006). These findings from animal experiments formed a valuable translational bridge to guide the subsequent clinical studies.

The results of three large-scale, randomized, double-blind, placebo-controlled clinical trials involving a total of 5,584 patients in Europe and the United States were reviewed by Carai et al. (2006). The trials involved daily treatment with 5 or 20 mg rimonabant or placebo for 1 year. Subjects were overweight or obese, and in addition to drug treatment, they agreed to observe a calorie-restricted diet. A variety of weight and other outcome measures were evaluated. In one of the trials, treatment was continued (or discontinued) for a further period of 1 year. The results were remarkable. All three studies yielded very similar data: In terms of weight loss after 1 year, patients receiving 20 mg rimonabant lost 6.3–6.9 kg compared to a loss of 1.5–1.8 kg in the placebo groups. The weight loss was accompanied by significant decreases in waist circumference, showing that the drug was particularly effective in reducing abdominal fat, which is known to be a risk factor for cardiovascular disease. In addition, rimonabant caused significant decreases in plasma glucose and fat levels, increases in plasma levels of adiponectin, reduced levels of leptin, and elevations in "good" high-density lipoprotein cholesterol—indicating protective effects against a number of known risk factors for heart disease. Patients maintained on rimonabant for a second year maintained the reduction in body weight and associated metabolic parameters.

Rimonabant appeared to be well tolerated and safe; the only side effects seen more frequently in patients receiving 20 mg rimonabant were episodes of dizziness, nausea, anxiety, and depression—but these occurred at very low frequencies. Given this promising background, rimonabant seemed set to be a "blockbuster" new medicine, with a wholly novel approach to treating obesity and metabolic disturbances. This was not to be: In 2007, the drug was removed from the European market due to suspected drug-induced suicides. It never gained FDA approval in the United States due to a low incidence of psychiatric effects (anxiety and depression). However, the effectiveness of rimonabant could be attributed in large part to its actions on peripheral CB-1 receptors in adipose tissue and liver. Given the promise that rimonabant seemed to have, it is not surprising that several competitive drugs were developed—for example, tiranabant (Merck), otenabant (Pfizer), surinabant (Sanofi), and ibipinabant (Solvay). The focus of interest more recently has been directed to the development of CB-1 antagonists deliberately designed to lack CNS penetration, in the expectation that actions on peripheral CB-1 sites would suffice to treat obesity and metabolic disturbances, although this concept remains unproven (Ward and Raffa, 2011). Preclinical reports of CB-1 antagonists lacking CNS penetration, however, show them to be effective in treating the symptoms of spasticity and body weight gain in animal models (LoVerne et al., 2009; Cluny et al., 2010; Alonso et al., 2012; Pryce et al., 2014).

Placebo Effects

The effects of homeopathic and herbal medicines and various other "alternative medicine" approaches most likely involve the well-documented placebo effect (Kaptchuk and Miller, 2015). If people are given a tablet or capsule that is identical to that containing a genuine medicine, but which contains no active ingredients other than sugar or some other inert powder, some patients will often report that they feel better. This even extends to the treatment of severe pain, where patients receiving placebo may report pain relief.

Some years ago, Jon Levine and Howard Fields, researchers at the University of California in San Francisco, conducted some ingenious experiments that shed some light on the mysterious placebo effect. They studied groups of students who had attended the student dental clinic for the surgical removal of wisdom teeth. The students were told that they would receive either an inactive placebo or the powerful painkiller morphine. Two hours after recovering from the anesthetic, the subjects who had volunteered to take part in the study received an intravenous injection of either morphine or a saline placebo. To ensure that the investigator did not inadvertently reveal whether the subjects were receiving morphine or placebo, the study was "blinded"—that is, neither the subjects nor the physician knew which subjects were receiving the active drug. This information was coded, and the code was broken only when the experiment was complete. After the dental operation, most people experienced pain, which increased gradually over a period of several hours. The subjects who received morphine reported that their pain was either stable or decreased. Those who received the placebo saline injection, however, fell into two groups. Approximately two-thirds of them showed no response, and their pain increased gradually over the course of the study, whereas approximately one-third of the placebo group were classified as "responders" because they reported pain relief that was equivalent to that of subjects who had received a moderate dose of morphine (Levine et al., 1979). In a subsequent study, it was found that the drug naloxone, which acts a potent antagonist of the actions of morphine at opiate receptors, could prevent the placebo response in "placebo responders," but it had no effect in "placebo non-responders." How could naloxone block the effect of a drug that the placebo group had not received? The answer seems to be that the mere expectation of pain relief from an injection that might contain morphine was by itself sufficient.

There is a real possibility that some of the medical benefits claimed by patients who are self-medicating with cannabis could lie in that category. The patients are usually those for whom conventional medicine has failed and they are turning to alternative medicine for relief from their symptoms. Cannabis has the added attraction to many of being a "natural" and "herbal" remedy, embedded in centuries of folklore and folk medicine. Currently, very few of the medical indications for which herbal cannabis is illegally used can be substantiated by data from scientifically controlled clinical trials. The thousands of patients who are currently self-medicating rely almost entirely on word-of-mouth anecdotal evidence and their own personal experiences of the drug. Anecdotal evidence, however, is not reliable and cannot be used to persuade regulatory agencies to approve cannabis as a medicine. To the non-scientist, this is difficult to understand. The often moving personal accounts of individuals who report the benefits they have derived from herbal cannabis are so compelling, what more is needed? Professor Grinspoon at Harvard has long been a passionate advocate of cannabis-based medicines and has given a fascinating series of accounts of patients' individual experiences (Grinspoon and Bakalar, 1993).

Summary

There are clearly several possible therapeutic indications for cannabis-based medicines, but for many of them, evidence for the clinical effectiveness of the drug is still inadequate by modern standards. The lists of medical indications for cannabis enshrined in state law in the United States include several that are not yet proven (e.g., Parkinson's disease, dementia, post-traumatic stress syndrome, Tourette's disease, and Crohn's disease). One of the obvious complications in the medical use of cannabis is that the "window" between its therapeutic effects and the cannabis-induced "high" is often narrow, making it difficult to conceal active drug treatment from placebo. As the Institute of Medicine (1999) report notes, however, this can sometimes be beneficial to the patient. Older patients with no previous experience of cannabis may find the psychic effects of the drug disturbing and unpleasant. But in some conditions, the anti-anxiety effects of cannabis can have a beneficial effect because anxiety itself tends to make the symptoms worse (e.g., in movement disorders, cancer chemotherapy, and AIDS wasting syndrome).

The Supply of Cannabis to "Cannabis Pharmacies"

The Dutch Model

The Netherlands has a long history of liberal drug policies. It was recognized, however, that the purchase of medical cannabis from "coffee shops" was not satisfactory: Patients did not know what they were getting, and their doctors did not know what they were prescribing. The Dutch government established the Office of Medicinal Cannabis (OMC) in 2000 to organize the provision of pharmaceutical-grade cannabis for medical use (Office of Medicinal Cannabis, Netherlands, 2016). A single grower was chosen—Bedrocan, which undertook the development of medical supplies of cannabis to pharmaceutical standards of growing, processing, and packaging. Sterility is provided by gamma-irradiation of the products. Quality is maintained by regular inspections by OMC. Bedrocan received approval by the European Union for its adherence to the strict rules of "good manufacturing practice," and it delivers a standardized medical product from plants grown in clean conditions, free of pesticides and fungal contamination. Bedrocan provides six different products covering a range of THC/cannabidiol contents:

Bedrocan: 22% THC/<1.0% cannabidiol (*Cannabis sativa*)
Bedrobinol: 13% THC/<1.0% cannabidiol (*C. sativa*)
Bediol: 6.3% THC/8.0% cannabidiol (*C. sativa*)
Bedica: 14% THC/<1.0% cannabidiol (*Cannabis indica*)
Bedropuur: 22% THC/<1.0% cannabidiol (*C. indica*)
Bedrolite: <10% THC/9.0% cannabidiol (*C. indica*)

Bedrocan can also supply placebo herbal material for research purposes, from which cannabinoids are extracted but flavorsome terpenes are replaced. Bedrocan products are exported to a number of other European countries to meet their requirements for medical-grade cannabis of consistent quality. The availability of placebo and cannabis products with standardized composition is a valuable resource for research.

An evaluation of the quality of medicinal-grade cannabis in the Netherlands involved the analysis of official samples obtained from OMC and comparison with cannabis samples obtained from randomly selected coffee shops. Several coffee shop samples were found to contain less weight than expected, and all were contaminated with bacteria and fungi. The results showed that medicinal cannabis obtained from pharmacies was more reliable and safer for the health of medical users (Hazecamp, 2006).

Canada

In April 2017, Canadian Prime Minister Trudeau announced that cannabis would be legalized in Canada beginning July 1, 2018. The legalization of cannabis in Canada involved a change in Canadian federal law, unlike the situation in the United States, in which cannabis remains illegal under federal law. In 2001, Canada was the first country to offer medical marijuana to those in need. After further refinements in August 2016, the Access to Cannabis for Medical Purposes Regulations (ACMPR) came into effect. This law provided for individual rights to cultivate a limited number of cannabis plants for individuals' own medical use and licensed a number of growers to provide pharmaceutical-grade cannabis to pharmacies for medical dispensing. In administering the ACMPR, the government health department Health Canada had two main roles: (1) licensing and overseeing the commercial industry and (2) registering individuals to produce a limited amount of cannabis for their own medical purposes (or to have another individual produce it for them). Since May 2017, Health Canada had licensed 45 growers, mostly commercial companies. They have to abide by the strict regulations laid down by Health Canada and are subject to routine inspections to confirm their compliance (Government of Canada, 2016; see http://www.hc-sc.gc.ca/dhp-mps/marihuana/info/list-eng.php). Regulations for growers include written standard operating procedures that cover facility, equipment, and production processes; a facility that is clean and controlled for sanitation and environmental conditions with an effective and monitored security system. A comprehensive inventory control system, as well as standardized packaging and labeling regulations. Once established, initial testing and ongoing testing for product integrity are required.

Several reputable growers have entered the Canadian market; for example, the Dutch company Bedrocan CV established a Canadian company, Bedrocan Canada Inc., and was allowed to import plant strains and expertise from the Netherlands. CanniMed, another large grower, offers three different cannabis oils and seven different herbal products, with a range of THC/cannabidiol content. Some growers offer plants for sale for licensed home cultivation. An analysis of the cannabinoid content of the products offered by 25 growers divided the products into "THC-dominant" ($n = 180$), "CBD-dominant" ($n = 67$), and "THC pure" ($n = 30$). Two-thirds of users preferred the THC-dominant products (THC content 7.44–29.1%). More than 90% of the THC-dominant products contained only traces of CBD. It was thought to be critical for the THC and CBD contents to be more balanced to reduce the risks of adverse side effects, and growers were encouraged to heed this advice, although no nationally required standards have been devised (Lewis, Yang, and Masilewski, 2017).

The United States

Although medical cannabis is sanctioned in many states, there are no nationally agreed upon rules to ensure quality control of the cannabis dispensed by cannabis pharmacies. Leafy, an influential source of information on cannabis, has reported a state-by-state analysis of the requirements of each state in which medical marijuana is legally available (Rough, 2016). Several states (Arizona, Michigan, Montana, and Rhode Island) and Washington, DC, have no requirements for product testing. The testing requirements for other states vary widely, with a majority requiring only freedom from contaminants (pesticides, microorganisms, molds, and solvents), whereas others (e.g., California and Colorado) have very detailed requirements, including a detailed cannabinoid profile. Some have seen this lacuna in the US medical cannabis field and have recommended national testing programs. In 2010, Steep Hill, a cannabis analysis laboratory in California, published a report on quality assurance for medical cannabis. The Association of Public Health Laboratories (2016) offered its comprehensive "Guidance for State Medical Cannabis Testing Programs."

Conclusion

There are clearly several possible therapeutic indications for cannabis-based medicines, but for many of them, evidence for the clinical effectiveness of the drug is still inadequate by modern standards. The lists of medical indications for cannabis enshrined in state law in the United States include several that are not yet proven (e.g., Parkinson's disease, dementia, post-traumatic stress syndrome, Tourette's disease, and Crohn's disease). One of the obvious complications in the medical use of cannabis is that the "window" between its therapeutic effects and the cannabis-induced "high" is often narrow, making it difficult to conceal active drug treatment from placebo. As the Institute of Medicine (1999) report noted, however, this can sometimes be beneficial to the patient. Older patients with no previous experience of cannabis may find the psychic effects of the drug disturbing and unpleasant. But in some conditions, the anti-anxiety effects of cannabis can have a beneficial effect because anxiety itself tends to make the symptoms worse (e.g., in movement disorders, cancer chemotherapy, and AIDS wasting syndrome). Placebo effects of cannabis may be important.

The other requirement for a human medicine is that it should be safe to use. Chapter 6 addresses whether this is the case for marijuana.

Chapter 6

Is Cannabis Safe?

The initial enthusiasm for cannabis in the 1960s and early 1970s was rapidly followed by a wave of reaction in the Western world. Many parents were appalled that their children were taking this relatively unknown drug and they feared that it might damage their health or impair their education. Although scientists are supposed to try to minimize bias, this has been difficult to avoid in a field so colored by issues of morality and public policy, and some have been guided by a moral commitment to prove that cannabis is harmful. Extravagant warnings were given, suggesting that cannabis was a highly dangerous drug that could cause chromosomal damage, impotence, sterility, respiratory damage, depressed immune system responses, personality changes, and permanent brain damage. These claims were later proved to be spurious, and the balanced reviews by Hollister (1986, 1992) and Zimmer and Morgan (1997) in their entertaining book *Marijuana Myths, Marijuana Facts* showed how effectively many of the claims could be demolished. It is thus not necessary to deal with all of these arguments in detail here but, rather, simply to highlight some of the factors that may determine whether cannabis is considered sufficiently safe to justify its reintroduction into the Western world as a legal drug or, alternatively, whether its overall prohibition remains justified. The electorate in some US states already decided, by a vote of greater than 50%, to legalize cannabis, although the geographical distribution of those in favor of legalization is uneven, with those favoring legalization concentrated in the Eastern and Western coastal areas (ProCon, 2016).

Safety Studies in Laboratory Animals

Delta-9-tetrahydrocannabinol (THC) is a very safe drug. Laboratory animals (rats, mice, dogs, and monkeys) can tolerate doses of up to 1000 mg/kg. This would be equivalent to a 70-kg person swallowing 70 g of the drug—about 5,000 times more than is required to produce a high. Long-term toxicology studies with THC were sponsored by the National Institute of Mental Health in the late 1960s (Braude, 1972). These included a 90-day study with a 30-day recovery period in both rats and monkeys. These studies were similar in design to those required for any new medicine before it can be approved for human use. Large numbers of animals were exposed to high doses of the drug every day, and blood samples were taken regularly to search for biochemical abnormalities during the study. At the end of the study, a careful autopsy was performed on each animal, recording the weight and appearance of internal organs. Sections of the major organs were subsequently examined under the microscope to look for any pathological changes. Interestingly, these studies included not only Δ^9-THC but also Δ^8-THC and a crude extract of marijuana. Treatment of animals with doses of cannabis or cannabinoids in the range 50–500 mg/kg led to decreased food intake and lower body weight. All three test substances initially depressed behavior, but later the animals became more active and were irritable or aggressive. At the end of the study, decreased organ weights were seen in ovary, uterus, prostate, and spleen, and increases were seen in adrenals. The behavioral and organ changes were similar in monkeys but less severe than those seen in rats. Further studies were carried out to assess the potential damage that might be done to the developing fetus by exposure to cannabis or cannabinoids during pregnancy.

Treatment of pregnant rabbits with THC at doses up to 5 mg/kg had no effect on birth weight and did not cause any abnormalities in the offspring. Braude concluded,

> In summary, I would like to say that Delta-9-THC given orally seems to be a rather safe compound in animals as well as in man and appears to have little teratological potential even at dose levels considerably higher than the typical human dose. (Braude, 1972, p. 99)

Chan et al. (1996) reported the findings of similarly detailed toxicology studies carried out with THC by the National Institute of Environmental Health Sciences in the United States, in response to a request from the National Cancer Institute. Groups of rats and mice were treated repeatedly with a range of doses of THC dissolved in corn oil, including doses many times higher than those likely to be used clinically. Each dose of the drug was administered to a separate group of 10 male and 10 female animals. In both species, the doses ranged from 0 to 500 mg/kg. The animals were treated five times a week for 13 weeks, and some groups of animals were followed for a further period of 9 weeks of recovery. By the end of the study, more than half of the rats treated with the highest dose (500 mg/kg) had died, but all of the remaining animals appeared healthy, although in both species the higher doses caused lethargy and increased aggressiveness. The THC-treated animals ate less food and their body weights were consequently significantly lower than those of untreated controls at the end of the treatment period, but they rose back to normal levels during the subsequent recovery period. During this period, animals were sensitive to touch and some exhibited convulsions. There was a tendency for the drug to cause decreases in the weight of the uterus and testes.

In further studies, groups of rats were treated with doses of THC up to 50 mg/kg and mice with doses up to 500 mg/kg, five times a week for 2 years, a standard test to determine whether new medical compounds are liable to cause cancers. At the end of the 2-year period, more treated animals had survived than controls—probably because the treated animals ate less and had lower body weights. The treated animals also showed a significantly lower incidence of the various cancers normally seen in aged rodents—in testes, pancreas, pituitary gland, mammary gland, liver, and uterus. Although there was an increased incidence of precancerous changes in the thyroid gland in both species and in the mouse ovary after one dose (125 mg/kg), these changes were not dose-related. The conclusion was that there was "no evidence of carcinogenic activity of THC at doses up to 50 mg/kg" (Chan et al., 1996). This was also supported by the failure to detect any genetic toxicity in other tests designed to identify drugs capable of causing chromosomal damage. For example, THC was negative in the Ames test, in which bacteria are exposed to very high concentrations of the test drug to determine whether it induces any mutations. In another standard test, hamster ovary cells were exposed to high concentrations of the drug in tissue culture, and no effects were observed on cell division that might indicate chromosomal damage.

There have been claims that chronic cannabis use may permanently damage the brain, but there is little scientific evidence to support this (for reviews, see Hollister, 1986, 1992; Zimmer and Morgan, 1997). The earlier studies have been complemented by the application of powerful modern neuroimaging methods. For example, a magnetic resonance imaging study compared 18 current, frequent, young adult cannabis users with 13 comparable non-users and found no evidence of cerebral atrophy or regional changes in tissue volumes (Block et al., 2000).

By any standards, THC must be considered to be a very safe drug both acutely and on long-term exposure. The very low lethality of the drug may reflect the fact that cannabinoid receptors are virtually absent from those regions at the base of the brain that are responsible for such vital functions as breathing and blood pressure control. The available animal data are more than adequate to justify its approval as a human medicine, and indeed it has been approved by the US Food and Drug Administration (FDA; as Marinol) for certain limited therapeutic indications.

Cannabis-Related Deaths

Despite the widespread illicit use of the cannabis, there are very few, if any, instances of people dying from an overdose. In Britain, the National Statistics Office listed no deaths related solely to cannabis in the period 2000–2004, while there were estimated to be 3 million cannabis users. The legalization debate clearly needs an answer to the question of whether cannabis can kill. This simple question, however, does not have any simple answer. It is maintained in the United States that there are no deaths primarily attributable to cannabis ingestion (see http://herb.co/2016/01/12/many-people-died-overdosing-marijuana;http://drugwarfacts.org/chapter/causes_of_death). Sidney et al. (1997) reviewed the topic and concluded, "In summary, this study showed little, if any, effect of marijuana use on non-AIDS mortality in men and on total mortality in women."(Sidney et al., 1917, p. 589). This accords with the general understanding that the number of cannabis-related deaths is close to zero. However, in the United Kingdom, the National Statistics Office (England and Wales) recorded 21 cannabis-related deaths in 2015 (28 in 2014, 11 in 2013, and 14 in 2012), and these included deaths in which alcohol and/or other drugs were a component. The National Programme on Substance Abuse Deaths (NPSAD, 2017) reported 21 cannabis-related deaths in England in 2014 (14 in 2013, 24 in 2012, and 14 in 2011). When the National Statistics Office considered deaths in which cannabis was the sole contributor, the numbers decreased to 4 in 2015, 7 in 2014, 1 in 2013, and 0 in 2012; this was also the case for NPSAD. The discrepancy between the United Kingdom and the United States is probably related to the definition of "drug-related deaths" (Family Council, 2016). Under a Freedom of Information suit, the FDA reported 279 deaths in which cannabis played a part during the period from January 1, 1997, to June 30, 2005. In the same time period, deaths from prescription drugs were far more numerous; for example, antipsychotics were reported as the prime cause of death in 1,593 cases and as a contributory cause in 702 (see https://www.oregon.gov/.../Marijuana/Public/DeathsFromMarijuanaV17FDAdrugs.pdf); this could possibly explain the discrepancy. By comparison with other commonly used drugs, these statistics are impressive. In the United States, there were a total of almost 30,000 deaths linked to prescription drug overdose, 17,536 of which involved prescription opioids (National Institute on Drug Abuse (NIDA), 2017). Even such apparently innocuous medicines as aspirin and related nonsteroidal anti-inflammatory compounds are not safe. It has been estimated that more than 100,000 Americans are hospitalized because of the tendency of these drugs to cause catastrophic gastric bleeding, and there are 16.500 deaths related to these drugs.

The NPSAD's 2017 annual report included a commentary on cannabis-related deaths and concluded, "While fatal overdose solely attributable to cannabis is rare, other forms of cannabis-related fatality are more common."

Acute Effects of Cannabis

Of all the immediate actions of cannabis (see Chapters 2 and 3), its psychoactive effects undoubtedly cause the greatest concern in considering the medical uses of the drug. In several of the medical applications that have been assessed to date, unwanted psychic side effects have been cited as the main reason for patients rejecting the drug as unacceptable. Patients who have had no prior experience with cannabis often find the intoxicant effects disturbing, and the drug may induce a frightening panic/anxiety attack in such people.

Others may simply not want to be "high" when they go about their daily work. The deleterious effects of cannabis on short-term memory and other aspects of cognition (see Chapter 3) make it especially unacceptable for those whose occupation depends on an ability to remain alert and capable of handling and processing complex information. If improved delivery systems could be devised, it is more likely that patients could self-adjust their optimum doses of the drug to avoid some of these unwanted effects, but it appears that the "therapeutic window" between a medically effective dose and an intoxicant one is narrow.

There are also quite profound effects of cannabis on the heart and vascular system. In inexperienced users, the drug can cause a large increase in heart rate (up to a doubling), and this could be harmful to someone with a previous history of coronary artery disease or heart failure. For this reason, such patients should be excluded from any clinical trials of cannabis-based medicines. The postural hypotension that can be caused by cannabis could also be distressing or possibly dangerous because the fall in blood pressure when rising from a seated or lying down position can result in fainting. The effects of the drug on the cardiovascular system usually show rapid tolerance on repeated exposure to cannabis, so for normal healthy subjects, these effects would not appear to be of any particular concern. Nevertheless, Yankey et al. (2017) studied mortality in a group of 1,213 US adults and concluded that although cannabis use did not increase the risk of death from heart disease, or cerebrovascular events (stroke), there was a significantly increased risk of death from a variety of causes linked to hypertension, particularly among long-term cannabis users. Unfortunately, the authors did not define the dose range used by "cannabis users."

The availability of "concentrates" with a high THC content can lead to overdosing, in which the user may collapse (Roberts, 2013). The misuse of potent forms of cannabis can clearly cause medical emergencies in both adults and children, and the accidental ingestion of cannabis-containing cookies and other edibles is an increasing problem (Family Council, 2016).

Can Cannabis Use Lead to Psychosis?

A serious acute reaction to cannabis is a form of toxic psychosis. The symptoms can be severe enough to lead to admission to emergency psychiatric wards. In some of the psychiatric literature, this is referred to as "cannabis psychosis" (or "marijuana psychosis") (Thomas, 1993; Castle and Murray, 2004). It nearly always results from taking large doses of the drug, often in food or drink, and the condition may persist for some days. The initial diagnosis can be confused with schizophrenia because the patients may display some of the characteristic symptoms of schizophrenic illness. These include delusions of control (being under the control of some outside being or force), grandiose identity, persecution, thought insertion, auditory hallucinations (hearing sounds, usually nonverbal in nature), changed perception, and blunting of the emotions. Not all symptoms will be seen in every patient, but there is a considerable similarity to paranoid schizophrenia. This has led some to propose a "cannabinoid hypothesis of schizophrenia," suggesting that the symptoms of schizophrenic illness might be caused by an abnormal overactivity of endogenous cannabinoid mechanisms in the brain (Emrich et al., 1997). Modern plant strains containing high- THC often contain very little cannabidiol, which exerts a modulating effect on the psychotomimetic effects of THC (Izzo et al., 2009; Niesink and van Laar, 2013; Iseger and Bossong, 2015). High THC-containing products, particularly concentrates, may be more likely to precipitate a temporary cannabis psychosis (Murray et al., 2016), although this is by no means proven.

Randomized, double-blind, placebo-controlled laboratory studies in human volunteers, using a variety of routes of administration, have demonstrated that cannabinoid agonists, including phytocannabinoids and synthetic agonists, produce a wide range of positive, negative, and cognitive symptoms in healthy volunteers (D'Souza et al., 2016; reviewed by Sherif et al., 2016). Several studies have shown that pretreatment with the non-psychoactive cannabinoid cannabidiol reduces the psychotomimetic effects of THC (Sherif et al., 2016). Schizophrenia-like psychotic symptoms can be induced in healthy volunteers within a few minutes following the intravenous injection of THC, and they persist for 1 or 2 hours (Morrison et al., 2009). Cannabis is not unique in sometimes causing acute psychotic reactions; similar effects are commonly seen with amphetamines, cocaine, ketamine, phencyclidine, and alcohol (Thirthalli and Benegal, 2006). Cannabinoid agonists have also been shown to exacerbate symptoms in patients with schizophrenia, and such individuals are also more vulnerable than healthy control subjects to the acute behavioral and cognitive effects of cannabinoid agonists (Sherif et al., 2016). Use of cannabis by patients with schizophrenia dose-dependently increases the risk of relapse (Schoeler et al., 2016).

The concern that the use of cannabis might precipitate mental illness in some users is long-standing. There was a lively correspondence in the columns of the *British Medical Journal* in 1893, for example, as to whether or not the endemic use of hashish in Egypt led to mania and insanity (*Br. Med. J.* 1893, pp. 710, 813, 920, 969, and 1027). There was concern that the mental asylums in British India were filling with cannabis-induced lunatics, and this was one of the factors that led the British government to appoint the Indian Hemp Drugs Commission (1894). The Commission undertook a large and painstaking review and concluded that there were virtually no patients in the Indian asylums whose illness could be attributed to cannabis use. The Commission's findings were not widely noted, however, and claims of a relationship between cannabis use and insanity continued to be made in India and in many other countries. Early advocates of marijuana prohibition in the United States used claims that cannabis use leads to insanity. In recent years, this debate has been reopened by the publication of a number of studies that show an association between teenage cannabis use and the development of psychotic symptoms in adulthood. Academic psychiatrists in Britain called on the government to reconsider the downgrading

TABLE 6.1 Crude and Adjusted Odds Ratios for Developing
Schizophrenia in Swedish Conscripts Cohort

Drug Use	No. of Subjects	No. Developing Schizophrenia	Crude Odds Ratio	Adjusted Odds Ratio[a]
No cannabis ever	36,429	215	1.0	1.0
Cannabis ever	5,391	73	2.2	1.5
Once	608	2	0.6	0.6
2–4 times	1,380	8	1.0	0.9
5–10 times	806	9	1.9	1.4
11–50 times	689	13	3.2	2.2
>50 times	731	28	6.7	3.1

[a]Adjusted odds ratio takes into account the following confounding factors: poor social
integration, disturbed behavior, cigarette smoking, and place of upbringing.

Source: Data from Zammit et al. (2002).

in the legal status of cannabis in light of these findings. The first study
making this claim was performed approximately 30 years ago. The
entire cohort of more than 50,000 male conscripts to the Swedish
army in 1969 was studied (Table 6.1). The conclusion was that
there was a dose-dependent relation between cannabis use by age
18 years and schizophrenia by age 45 years, with a threefold increase
in risk for those reporting using cannabis 50 times by age 18 years
(Andreasson et al., 1987).

Similar longitudinal studies have been reported in several other
countries, and the 10 studies of this type now available all confirm the
association between early use of cannabis and subsequent psychotic
illness (reviewed by Gage et al., 2016). There seems little doubt that
an association between cannabis use and subsequent schizophrenia
or other psychotic illness does exist. In a meta-analysis of the studies
then available, Moore et al. (2007) estimated a 40% increased risk
of developing psychotic illness in cannabis users compared to those
who had never taken cannabis, and this was updated to an increased

risk of 45%, taking additional studies into account (Gage et al., 2016). One study suggested a dose–response relationship between low- and high-potency cannabis use, with the latter leading to first episode of psychosis 6 years before that of people who never used cannabis; Murray et al. (2016) argued that the increasing potency of cannabis and the availability of potent synthetic cannabinoids increased the risk of psychosis. The association between teenage cannabis use and subsequent psychotic illness could be explained by some common vulnerability—for example, a genetic disposition to cannabis and risk of psychosis. This view is taken and apparently vindicated by the finding of novel cannabis-dependence risk alleles potentially linked to psychotic illness (Sherva et al., 2016) and that familial liability to psychosis is expressed as differential sensitivity to cannabis (Korver et al., 2011). Genetic variations in the *ACT1* gene (a member of three closely related protein kinases) were related to increased psychotomimetic effects of smoked marijuana (Morgan et al., 2016b). Research on the genetics of schizophrenia has increased rapidly in the past few years (Giegling et al., 2017). Modern techniques for whole genome studies have been applied in the search for cannabis-dependence genes, with some success (Volkow et al., 2016; Gizer et al., 2018). Although the existence of causal relationship has been criticized (Kair and Hart, 2016), others continue to argue that a causal relationship is likely and that the evidence provided so far on shared vulnerability factors is inadequate (Murray and Di Forti, 2016; Murray et al., 2016). D'Souza et al. (2016) take an in-between stance, arguing that cannabis use may be a "component cause" of later psychosis.

The question of whether the research on cannabis and schizophrenia warrants a public health warning has been advocated (Murray and Di Forti, 2016; Murray et al., 2016). However, the data suggest that only 2% of cannabis users are at risk of psychotic illness and that 5,000–10,000 heavy cannabis users would have to stop to prevent a single case of schizophrenia (Gage et al., 2016). These figures suggest that a public health warning would be unlikely to attract much support. Nevertheless, this field of research is active, and a better understanding will result—this is certainly an important issue for those treating young cannabis users.

Effects of Cannabis on Motivation

As early as the late 19th century, reports from India suggested that heavy cannabis use was associated with apathy, defined as reduced motivation for goal-directed behavior (Indian Hemp Drugs Commission, 1894). However, it was not until the 1960s that the amotivational effects of chronic cannabis use were linked to impairments in learning and sustained attention. The term *cannabis amotivational syndrome* was proposed by McGlothlin and West (1968), who characterized it as apathy and diminished ability to concentrate, follow routines, or successfully master new material. There is both preclinical and clinical evidence supporting the view that cannabis use is associated with an amotivational state. In rhesus monkeys, heavy chronic cannabis use or administration has been found to dampen motivation, as measured on operant tests (Paule et al., 1992), and laboratory evidence has been reported supporting an association between reduced motivation for reward-related behavior in cannabis users compared with control individuals (Lane et al., 2005). Lawn et al. (2014) found an acute effect of cannabis with or without cannabidiol in reducing performance in a computer test for money. The same test in subjects who were dependent on cannabis did not give clear-cut results. A possible underlying cause of the acute amotivational effects is suggested by reports that cannabis users exhibit reduced striatal dopamine synthesis capacity, with an inverse relationship to amotivation (Bloomfield et al., 2014). As with cognitive effects, cannabis use may be a cause, a consequence, or a correlate of altered motivation. (Volkow et al., 2016).

Effects of Cannabis on Driving

Animal experiments have shown that THC has characteristic effects on the ability to maintain normal balance; movements become "clumsy," and at higher doses the animals maintain abnormal postures and may remain immobile for considerable periods (Adams and Martin, 1996). Marijuana similarly affects human subjects, impairing their performance in tests of balance and reducing their performance in tests that require fine psychomotor control (e.g., tracking a moving point of light on a screen with a stylus) or manual dexterity (for review, see Iversen, 2003). There is a tendency to slower reaction times. In these respects, marijuana has similar effects to those observed with intoxicating doses of alcohol. An obvious concern is whether these impairments make it unsafe for marijuana users to drive while intoxicated. Driving not only requires a series of motor skills but also involves a complex series of perceptual and cognitive functions. There have been numerous studies in which the effects of marijuana have been assessed on performance in "driving simulators" and even a few studies that were conducted in city traffic. Meta-analyses of multiple studies found that the risk of being involved in an accident significantly increased after marijuana use; in a few cases, the risk doubled or more than doubled (Asbridge et al., 2012; NIDA, 2016a). However, a large case–control study conducted by the US National Highway Traffic Safety Administration found no significant increased crash risk attributable to cannabis after controlling for drivers' age, gender, race, and the presence of alcohol (Compton and Berning, 2015). Several of the early studies showed no impairments at all, but as the driving simulators became more sophisticated and the tasks required more complex and demanding actions, impairments were observed, for example, in peripheral vision and lane control. Marijuana users, however, seem to be aware that their driving skills may be impaired, and they tend to compensate by driving more slowly, keeping some distance away from the vehicle ahead of them, and in general taking fewer risks (Smiley, 1986). This is in marked contrast to the effects of alcohol, which produces clear impairments in many aspects of driving skill as assessed in driving simulators. Alcohol also tends to encourage people to take greater risks and to drive more aggressively. There is no question that alcohol is a major contributory factor to road traffic accidents; it is implicated in as many as half of all fatal road traffic accidents. Nevertheless, driving while under the influence of cannabis cannot be recommended as safe. It is worth considering that the early studies were done in an era of relatively low-potency cannabis; with modern form of "skunk" and "concentrates," users may become more highly intoxicated and their driving skills more severely impaired. Studies in North America and in Europe that found that as many as 10% of drivers involved in fatal accidents tested positive for THC, but were complicated by the fact that in a majority of these cases (70–90%), alcohol was detected as well. It may be that the greatest risk of cannabis in this context is to amplify the impairments caused by alcohol when as often happens both drugs are taken together (Robbe, 1998).

According to federal data, auto accident fatalities have declined significantly during the past two decades—during the same time that a majority of US states have legalized marijuana for either medical or social use. In 1996, when California became the first state to legalize medical marijuana, the National Highway Traffic Safety Administration reported that there were an estimated 37,500 fatal car crashes on US roadways. This total declined to less than 30,000 by 2014. A recently published analysis reported that 3 years after recreational marijuana legalization, changes in motor vehicle crash fatality rates for Washington and Colorado were not statistically different from those in similar states without recreational marijuana legalization (Aydelotte et al., 2017).

Along with the psychic effects of cannabis are impairments in psychomotor skills so that the ability of users to carry out any tasks that require manual dexterity is likely to be impaired (see

Chapter 3). A drug-induced impairment of balance could also make elderly patients more likely to fall. A comparison of 452 marijuana smokers with a similar number of non-smokers attending the Kaiser Permanente health group in California revealed that the marijuana smokers had an increased risk of attending outpatient clinics with injuries of various types—perhaps as a result of the acute intoxicant effects of the drug (Polen et al., 1993).

Are There Persistent Cognitive Deficits?

The acute and long-term effects of marijuana on cognitive function were reviewed in Chapter 3. The acute impairment of working memory is relatively short-lived, disappearing after 3 or 4 hours as the marijuana high wears off. However, as described in Chapter 3, cognitive deficits can be detected during abstinence, for up to 20 days, and some impairments persist for 3 weeks or longer after cannabis use ends. Considerable attention has been paid to the possibility that people who use large doses of marijuana regularly may suffer long-term cognitive impairment. Most recent analyses of the literature have concluded that most of these deficits are reversible with prolonged abstinence, but some reports claim that impairment of verbal memory may be more persistent (Crean et al., 2011; Broyd et al., 2016), whereas others have failed to find any deficit in performance on neuropsychological tests when compared with non-users, after correction for confounding factors (Schreiner and Dunn, 2012). As Volkow et al. (2016) note, the effects of cannabis may be causal but may alternatively reflect genetic differences in susceptibility to cannabis.

Vulnerability of the Teenage Brain

There is increasing evidence that the adolescent brain may be more vulnerable than the adult brain to adverse effects of cannabis.

Endocannabinoids are involved in key neurodevelopment processes, so introducing exogenous cannabinoids may disrupt brain development (Maccarrone et al., 2014; Lubman et al., 2015). Preclinical research supports this view. Adolescent rats treated with a cannabinoid showed persistent deficits in object recognition tasks, whereas adult rats treated in the same way showed no persistent deficits (Schneider et al., 2008). Several studies have found that early onset of cannabis use is associated with greater neuropsychological impairment (Fontes et al., 2011; Gruber et al., 2012). Neuroimaging studies in adolescent- and adult-onset cannabis users have yielded conflicting results: The conclusion seems to be that cannabis use is not associated with any structural alterations in the brains of adolescent- or adult-onset users (Weiland et al., 2015), although cannabis users may have reduced neural connectivity (Zalesky et al., 2012; Houck et al., 2013; Orr et al., 2013). Volkow et al. (2016) caution that a number of confounding factors may have influenced these results, and a cause-and-effect relationship between teenage cannabis use and neuropsychological deficits remains unproven. Nevertheless, the possibility that teenage use of cannabis is particularly dangerous has been widely used in messages about the risks of cannabis use. Those younger than age 18 years in Europe and those younger than age 21 years in the United States are "under the radar" of both medical and legal recreational marijuana programs. Teenagers will continue to access and use cannabis of dubious quality from illegal dealers.

Tolerance and Dependence

For a review of tolerance and dependence, see Ramesh et al. (2011).

Many drugs when given repeatedly tend to become less effective so that larger doses have to be given to achieve the same effect—that is, *tolerance* develops. There are many examples of tolerance to THC and other cannabinoids in animals treated repeatedly with these drugs. Tolerance can be seen in experimental animals even after treatment with quite modest doses of THC, but it is most profound when large doses (>5 mg/kg) are employed. With very high doses (as much as 20 mg/kg per day), animals may become almost completely insensitive to further treatment with THC. When animals become tolerant to THC, they also demonstrate cross-tolerance to any of the other cannabinoids, including the synthetic compounds WIN-55,212-2 and CP-55,940 (Iversen, 2003). This suggests that the mechanism underlying the development of tolerance has something to do with the sensitivity of the cannabinoid receptors or some mechanism downstream of these receptors rather than simply to a more rapid metabolism or elimination of the THC. Repeated treatment with THC in both animals and people does tend to lead to an increased rate of metabolism of the drug—probably because drug-metabolizing enzymes in the liver are induced by repeated exposure to the drug. However, these changes are not large enough to explain the much larger changes in sensitivity seen in responses to the drug, including reduced effects on the cardiovascular system, body temperature, and behavioral responses. A more likely explanation is that repeated exposure to high doses of THC leads to a compensatory decrease in the sensitivity or number of cannabinoid receptors in brain. Several studies have reported decreases in the density of CB-1 receptor binding sites in the brains of rats treated for 2 weeks with high doses of THC or CP-55,940, although the molecular mechanisms are not clear.

In human volunteers exposed repeatedly to large doses of THC under laboratory conditions, tolerance to the cardiovascular and psychic effects can be produced as in the animal studies. However, it is not clear that tolerance occurs to any significant extent in people who use modest amounts of marijuana (Earleywine, 2002). Casual users taking the drug infrequently or those using small amounts for medical purposes seem to develop little, if any, tolerance. Patients in clinical trials of cannabis-based medicines maintain a constant dose for periods of more than 1 year (see Chapter 5). Tolerance seems most likely to become important for heavy users who are taking gram quantities of cannabis on a daily basis.

The question of whether regular users become "dependent" on the drug has proved to be one of the most contentious in the field of cannabis research. Those opposed to the use of marijuana believe that it is a dangerous drug of addiction, by which young people can easily become hooked. On the other hand, proponents of cannabis claim that it does not cause addiction and dependence at all, and users can stop at any time of the own free will. Abuse and addiction have been defined and redefined by various organizations throughout the years. The most influential current system of diagnosis is that published by the American Psychiatric Association (2015) in the fifth edition of the *Diagnostic and Statistical Manual of Mental Disorders* (DSM-V). The DSM-V uses the term "cannabis use disorder" (Martin et al., 2008; Hasin et al., 2013) and defines this as a cluster of symptoms indicating that the individual continues to use the substance despite significant substance-related problems. DSM-V differs from DSM-IV in considering "drug use disorder" as a spectrum ranging from substance abuse to substance dependence (Box 6.1).

The DSM-V defines a number of drug-related disorders, of which cannabis is only one of nine categories. DSM-V also defines "cannabis withdrawal" for the first time, a subject that has been controversial because there are few physical symptoms of withdrawal. Nevertheless, the symptoms described are seen consistently in heavy drug users after cannabis withdrawal (Budney et al., 2004; Preuss et al., 2010; Bonnet and Preuss, 2017). The DSM-V criteria for diagnosing withdrawal are summarized in Box 6.2.

This way of thinking about drug dependence is significantly different from much of the earlier work in this field. It means that neither tolerance nor physical dependence need necessarily be present to make the diagnosis of cannabis use disorder. This has particularly changed the way in which cannabis is viewed. It has often been argued that because tolerance and physical dependence are not

BOX 6.1 Diagnosis of Cannabis Use Disorder—DSM-V

A problematic pattern of cannabis use leading to clinically significant impairment or distress, as manifested by at least two of the following, occurring within a 12-month period:

- Cannabis is often taken in larger amounts or over a longer period than was intended.
- There is a persistent desire or unsuccessful efforts to cut down or control cannabis use.
- A great deal of time is spent in activities necessary to obtain cannabis, use cannabis, or recover from its effects.
- Craving, or a strong desire or urge to use cannabis.
- Recurrent cannabis use resulting in a failure to fulfill major role obligations at work, school, or home.
- Continued cannabis use despite having persistent or recurrent social or interpersonal problems caused or exacerbated by the effects of cannabis.
- Important social, occupational, or recreational activities are given up or reduced because of cannabis use.
- Recurrent cannabis use in situations in which it is physically hazardous.
- Cannabis use is continued despite knowledge of having a persistent or recurrent physical or psychological problem that is likely to have been caused or exacerbated by cannabis.
- Tolerance, as defined by either a (1) need for markedly increased cannabis to achieve intoxication or desired effect or (2) markedly diminished effect with continued use of the same amount of the substance.
- Withdrawal, as manifested by either (1) the characteristic withdrawal syndrome for cannabis or (2) cannabis is taken to relieve or avoid withdrawal symptoms.

Source: Reprinted with permission from the *Diagnostic and Statistical Manual of Mental Disorders*, Fifth Edition (Copyright © 2013). American Psychiatric Association. All Rights Reserved.

BOX 6.2 Diagnosis of Cannabis Withdrawal—DSM-V

Cessation of cannabis use that has been heavy and prolonged (i.e., usually daily or almost daily use over a period of at least a few months).

- Three or more of the following signs and symptoms develop within approximately 1 week after cessation of heavy, prolonged use:
  - Irritability, anger or aggression
  - Nervousness or anxiety
  - Sleep difficulty (i.e., insomnia, disturbing dreams)
  - Decreased appetite or weight loss
  - Restlessness
  - Depressed mood
  - At least one of the following physical symptoms causing significant discomfort: abdominal pain, shakiness/tremors, sweating, fever, chills, or headache
- The signs or symptoms cause clinically significant distress or impairment in social, occupational, or other important areas of functioning.
- The signs or symptoms are not attributable to another medical condition and are not better explained by another mental disorder, including intoxication or withdrawal from another substance.

Source: Reprinted with permission from the *Diagnostic and Statistical Manual of Mental Disorders*, Fifth Edition (Copyright © 2013). American Psychiatric Association. All Rights Reserved.

prominent features of regular marijuana users, the drug cannot be addictive. The DSM-V definition of cannabis use disorder is made as the result of a carefully structured interview, and the diagnosis rests on the presence or absence of various items from a checklist of symptoms. When such assessments are made on groups of regular marijuana users, a surprisingly high proportion are diagnosed positively. The National Epidemiological Survey on Alcohol and Related Conditions 2012–2013 reported that of 36,309 subjects aged 18 years or older included in the survey, 2.5% had experienced cannabis use disorder, as defined by DSM-V, in the past 12 months, and 6.3% had experienced it in their lifetime (Hasin et al., 2016). These are staggering numbers, implying that 6 million Americans had experienced cannabis use disorder in the past 12 months. Cannabis use disorder was often accompanied by other substance use disorders, depression, anxiety, and personality disorders; persistent cannabis use disorder was associated with disability (Hasin et al., 2016). As the number of Americans reporting cannabis use has doubled in the past decade, the number experiencing cannabis use disorder has also escalated (Bergland, 2016; Science Daily, 2016). The medical users of the drug usually take relatively small doses of cannabis on an intermittent basis and are therefore much less likely to become dependent. Case reports from individual patients often stress that they do not want to become high and that they use the drug only occasionally. Data from the controlled trials of Sativex show that even when treated for 2 years or more, patients did not increase the dose used, which remained surprisingly constant.

There have also been developments in basic animal research that point to similarities between cannabis and other drugs of addiction (Gonzalez et al., 2005). The availability of rimonabant, for example, has shown that physical dependence accompanied by a withdrawal syndrome can be seen in animals that have been treated for some time repeatedly with THC or other cannabinoid when they are challenged with the antagonist drug. The withdrawal signs in rats include "wet dog shakes" (a characteristic convulsive shaking of the body as though the animal's fur is wet—a behavior also seen typically during opiate withdrawal), scratching and rubbing of the face, compulsive grooming, arched back, headshakes, spasms, and backwards walking. In dogs, the withdrawal signs include withdrawal from human contact, restlessness, shaking and trembling, vomiting, diarrhea, and excess salivation. The reason why such withdrawal signs are not normally seen in animals or in people when cannabinoid administration is suddenly stopped is probably related to the long half-life of THC and some of its active metabolites in the body. This means that the CB-1 receptor is still exposed to low levels of cannabinoid for some time after the drug is stopped. With the antagonist drug, however, the CB-1 receptor is suddenly blocked.

The animal findings with a rimonabant challenge have an interesting parallel with research on the benzodiazepine tranquillizers, of which Valium (diazepam) is the best known example. These, too, were thought not be addictive because there was little evidence for any withdrawal syndrome on terminating drug treatment. However, when the first benzodiazepine receptor antagonist drug, flumazenil, became available, it soon became clear that withdrawal signs could be precipitated in drug-treated animals when challenged with this antagonist. As with THC, the benzodiazepines persist for long periods in the body, so drug withdrawal can never be abrupt. It is now generally recognized that benzodiazepine tranquillizers and sleeping pills do carry a significant risk of dependence on repeated use.

One way in which scientists can assess the addictive potential of psychoactive drugs is to determine whether animals can be trained to self-administer them. Rats, mice, or monkeys easily learn self-administration of heroin or cocaine. Indeed, rats will self-administer cocaine to the exclusion of all other behavior, including feeding and sex. They have to be given restricted access to the drug to avoid damaging their health. It has proved much more difficult to train animals to self-administer THC, however, and this fact has often been used to argue that THC has no addictive liability. In a key study, THC was dissolved in a vehicle containing 0.4–1.0% Tween-80 and 0.4–1.0%

ethanol in saline and injected intravenously in squirrel monkeys using low doses to replicate those taken by human users. When THC was preceded by a training stimulus of cocaine, monkeys rapidly learned to self-administer the drug (Tanda et al., 2000). Subsequently, it was shown that self-administration of THC could be demonstrated in drug-naive squirrel monkeys with an optimum dose of 4 µg/kg (Justinova et al., 2003,2005). These were important milestones, showing that the active ingredient of cannabis was rewarding and consequently self-administered by drug-naive animals.

Another series of experiments in animals revealed that in common with other drugs of addiction, THC is able to selectively activate nerve cells in the brain that contain the chemical transmitter dopamine (Oleson and Cheer, 2012). French et al. (1997) first reported that small doses of THC activated the electrical discharge of dopamine-containing nerve cells in the ventral tegmentum region of rat brain; they recorded this discharge electrically with microelectrodes (Figure 6.1). Tanda et al. (1997) subsequently confirmed this finding by direct measurements of dopamine release from the nucleus accumbens region of the rat brain, which contains the terminals of the nerves originating from the ventral tegmentum. They perfected a delicate technique that involves the insertion of minute probes into this region of rat brain, through which chemicals released in the brain can be monitored continuously in conscious freely moving animals (a method known as microdialysis). Earlier work from this group and a number of other laboratories had shown that a number of drugs of addiction selectively activate dopamine release in this region of the brain; the drugs included heroin, cocaine, *d*-amphetamine, and nicotine. To this list, they now include THC, adding to speculation about its status as a drug of "addiction" (Figure 6.1).

Subsequent research showed that synthetic cannabinoids can also promote dopamine release, using doses much lower than those used with THC (Oleson and Cheer, 2012; De Luca et al., 2015; ElSohly and Gul, 2015). The actions of cannabinoids were blocked

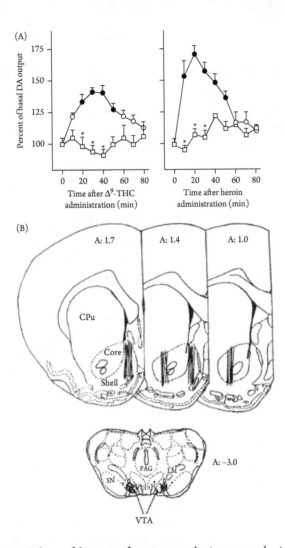

FIGURE 6.1 Release of dopamine from intact rat brain measured using microdialysis probes. (A) Dopamine release is stimulated by the administration of THC (0.15 mg/kg IV) or heroin (0.03 mg/kg IV) (circles). Solid circles indicate data points that were significantly different from baseline control values. In animals treated with the opiate M receptor antagonist naloxonazine, neither THC nor heroin any longer caused dopamine release (squares). (B) Sections of rat brain drawn to indicate the positions of the microdialysis probes in the individual animals used. Core, core of nucleus accumbens; Cpu, caudate putamen; Shell, shell of nucleus accumbens; SN, substantia nigra; VTA, ventral tegmentum; PAG = periaqueductal grey. On each section, "A" indicates the anterior coordinate, measured in millimeters from bregma.

Source: From Tanda et al. (1997).

by rimonabant and were absent in CB-1 receptor knockout animals (De Luca et al., 2015). Furthermore, the THC-induced release of dopamine seemed to involve an opioid mechanism because the effect of THC could be prevented by treatment of the animals with naloxonazine, a drug that potently and selectively blocks opioid receptor sites in brain. These results thus suggested that THC acts in part by promoting the release of opioid peptides in certain regions of the brain and that one of the consequences of this is to cause an increase in dopamine release in the nucleus accumbens. The precise biological meaning of this remains unclear. Most scientists do not believe that dopamine release per se explains the pleasurable effects of drugs of addiction, but it does seem to have some relation to whether the animal or person will seek to obtain further doses of that drug. Dopamine release in the nucleus accumbens is triggered by a variety of stimuli that are of significance to the animal, including food and sex. The ability of THC to activate opioid mechanisms also does not mean that THC is equivalent to heroin. Clearly, animals and humans can readily distinguish the distinct subjective experiences elicited by the two drugs, and THC or other cannabinoids do not mimic the severe physical dependence and withdrawal signs associated with chronic heroin use. Nevertheless, there is growing evidence that the naturally occurring opioid and cannabinoid systems represent parallel and sometimes overlapping mechanisms. Rats made dependent on heroin and then challenged with the opiate antagonist naloxone exhibit a strong withdrawal syndrome, with various characteristic behavioral features—for example, "wet dog shakes," teeth chattering, writhing, jumping, and diarrhea. Interestingly, some of these features are seen in a milder form if heroin-dependent animals are challenged with rimonabant. Conversely, rats treated repeatedly with high doses of cannabinoids will exhibit mild signs of withdrawal when challenged with the opiate antagonist naloxone. More support for the concept of a link between the cannabinoid and opioid systems in brain has come from CB-1 receptor knockout mice (Ledent et al., 1999; reviewed by Valverde et al., 2005). These animals survive quite normally without the CB-1 receptor, but as expected, they are unable to show any of the normal central nervous system responses to THC (analgesia, sedation, and hypothermia). Interestingly, the mice are also less responsive to morphine. Although morphine is still analgesic, it is less likely to be self-administered, and the mice display a milder opiate withdrawal syndrome. Further support for the existence of a genuine cannabis withdrawal syndrome in animals came from De Fonseca et al. (1997), who reported that there were elevated levels of the stress-related chemical corticotropin-releasing factor (CRF) in rat brain when rats were withdrawn from treatment for 2 weeks with the potent cannabinoid HU-210. Elevated levels of brain CRF were also seen in animals during withdrawal from alcohol, cocaine, and heroin. DSM-V also defines "cannabis withdrawal" for the first time, a subject that has been controversial because there are few physical symptoms of withdrawal. Nevertheless, the symptoms described are seen consistently after cannabis withdrawal from heavy drug users (Budney et al., 2004; Preuss et al., 2010; Bonnet and Preuss, 2017).

How Likely Are Recreational Cannabis Users to Develop Cannabis Use Disorder?

For a review of the information discussed in this section, see Roffman and Stephens (2006).

Wayne Hall and Nadia Solowij (1997), internationally recognized experts in the field of addiction research, described how they viewed the situation at the end of the 20th century:

Dependence on cannabis is the most prevalent and under-appreciated risk of regular cannabis users. About 10% of those who ever use cannabis, and one third to one half of those who use it daily will lose control over their cannabis use and continue to use the drug in the face of problems they believe are caused or exacerbated by its use. . . . Uncertainty remains as to how difficult it is to overcome cannabis dependence and what is the best way to assist individuals to become abstinent.

In the United States, cannabis use disorder is said to be increasing rapidly: "DSM-5 cannabis use disorder is a highly prevalent, co-morbid, disabling disorder that commonly goes untreated" (Hasin et al., 2016). For some people, the drug will come to dominate their lives. They will feel a psychological need and craving for the drug, and they will become preoccupied with locating continuing supplies of the drug. Consumption of marijuana may become so frequent that the user remains almost permanently stoned. They may prepare a joint before going to sleep at night in order to ensure that it is available for the morning. The severely dependent user is permanently cognitively impaired, lacks motivation, tends to suffer from lowered self-esteem, and may be depressed and is unlikely to be able to function at all in work or education. Although most regular cannabis users suffer merely mild discomfort when they stop taking the drug, the severely dependent user will suffer a definite syndrome of unpleasant withdrawal symptoms, as described by DSM-V. Cannabis use disorder is still largely unrecognized because many users continue to believe that it is not an addictive drug. There is a real need to educate cannabis users—to convey the message that they do run a risk of allowing the drug to dominate their lives.

The DSM-V definition of cannabis use disorder is not only a label, but it allows for a gradation of increasingly disabling symptoms. In reality, there are many degrees of dependence, as shown clearly by a detailed analysis of the wide variety of symptoms reported by 1,474 cannabis users in the United States (Denton and Earlywine, 2006). Hasin et al. (2016) found that among more than 4,584 users who described having experienced cannabis use disorder, 2,242 reported any occurrence, 1,002 a mild form, and 1,240 moderate to severe forms of cannabis use disorder. Among the whole population of cannabis users, there is probably a continuous gradation from harmless weekend users to heavy users whose severe dependence may wreck their lives.

If one attempts to assess the risk of dependence by comparison to other addictive drugs, cannabis does not score top of the list in terms of either the severity of the addiction or the likelihood of becoming hooked. The Institute of Medicine (1999) suggested that 9% of those who ever used cannabis become dependent (as defined by the DSM-IV criteria), compared to estimated dependency risks of 32% for tobacco, 23% for heroin, 17% for cocaine, and 15% for alcohol. Cocaine and heroin are far more damaging in terms of both the severity of the physical withdrawal syndrome that users will experience if they stop taking the drug and the probability of becoming addicted to the drug. Nicotine is notorious in the sense that a very high proportion of cigarette smokers tend to become permanent smokers after consuming only a few packs of cigarettes (Kozlowski et al., 1989). Unlike the casual user of marijuana, the cigarette smoker typically smokes 15–20 cigarettes a day, every day of the year. Unlike cigarette smokers, most marijuana smokers also seem to be able to give up the habit relatively easily. As they reach their 30s and become responsible for a family and a job, many are no longer willing to take the risk of being punished for illegal drug use.

Another way of measuring the extent of cannabis dependence is to ask how many people seek treatment for it. In Britain, the latest figures showed 18,3454 new entrants for cannabis addiction treatment, approximately double the number 10 years ago (European Monitoring Centre for Drugs and Drug Addiction, 2017). In the United States, Hasin et al. (2016), in the first national survey since the introduction of the DSM-V, reported that cannabis use disorder was common and largely untreated. These are alarming statistics, which at first sight suggest that cannabis dependence is a fast-growing problem, perhaps in part as a consequence of the increased use of very high-potency cannabis products. This may be partly true, but there are other factors at work. In some countries, notably the

United States and Canada, treatment is frequently offered as an alternative to criminal punishment for minor possession offenses or as a result of testing positive in the workplace. There is little doubt that cannabis use disorder is a problem and is likely to become more prevalent. Unfortunately, there have been very few controlled trials to assess treatment methods, which remain mainly focused on group counseling.

Is Cannabis a "Gateway Drug"?

A widely debated question is whether the use of cannabis leads people to use other illicit drugs and eventually to become addicted to these. Those who believe this to be true argue that even if marijuana is a relatively harmless drug, it can act as a "stepping stone" to other far more dangerous drugs (NIDA, 2017). This is a difficult question to address scientifically. Many surveys have shown that young people who use illegal psychoactive drugs begin with alcohol and tobacco and then marijuana. They tend to experiment with a number of other illicit drugs. Most who take heroin or cocaine will have had previous experience with marijuana and several other illicit substances. Kandel and Davies (1996), for example, surveyed 7,611 students aged 13–18 years in 53 New York state schools. Of the total, 995 had experience with marijuana and 403 had experience with cocaine, 121 of whom had taken crack cocaine. Alcohol and/or cigarette use tended to begin at age 12 or 13 years, marijuana use at age 15 years, and cocaine use at age 15 or 16 years. The young people who used drugs lived in social environments in which they perceived the use of drugs to be prevalent. Of the students who used crack cocaine, two-thirds reported that all or most of their friends had used marijuana and 38% had used cocaine. Among non-users of drugs, the corresponding figures were 8% and 0%, respectively. However, this does not prove that one drug leads to another, as Zimmer and Morgan (1997) note:

In the end, the gateway theory is not a theory at all. It is a description of the typical sequence in which multiple drug users initiate the use of high-prevalence and low-prevalence drugs. A similar statistical relationship exists between other kinds of common and uncommon related activities. For example, most people who ride a motorcycle (a fairly rare activity) have ridden a bicycle (a fairly common activity). Indeed the prevalence of motorcycle riding among people who have never ridden a bicycle is probably extremely low. However, bicycle riding does not cause motorcycle riding, and increases in the former will not lead automatically to increases in the latter. Nor will increases in marijuana use automatically lead to increases in the use of cocaine or other drugs.

Kandel and Davies (1996) found that parental behavior was an important determinant of the drug users' behavior. Parental use of alcohol and cigarettes was important in determining experimentation with these drugs. Perhaps more surprisingly, parental use of a medically prescribed tranquillizer was likely to be associated with children's experimentation with illicit drugs. Through their use of legally available psychotropic drugs, parents may indicate to their children that drugs can be used to handle their own feelings of psychological distress. So is the relationship that does exist between marijuana use and "harder" drugs simply a matter of social context? Is it the introduction to the underground world of illicit drugs through the black market in marijuana that leads people to experiment with other illicit substances? The Dutch believe that separating the supply of "hard" drugs from that of marijuana, and making the latter freely available, can break this relationship:

> It seems that Holland can justly claim to have separated the heroin and cannabis markets. As a result, young people are far less likely in Holland than elsewhere to experiment with heroin. Although there is room for argument over precisely how this has been

achieved, it is difficult to deny that the policy of separation of markets, including the toleration of coffee shops, has made a contribution. . . . By doing so the Dutch have provided persuasive evidence against the gateway theory of cannabis use, and in favour of the theory that if there is a gateway it is the illegal market place. (Zimmer and Morgan, 1997)

Is there any scientific basis for the gateway theory? Basic research has shown that THC can trigger activity in neural pathways in animal brain that use the chemical messenger dopamine (Tanda et al., 1997). The significance of this finding is that this is a common feature seen in response to a variety of addictive central nervous system drugs, including alcohol, nicotine, cocaine, amphetamines, and heroin. Some scientists have argued that it is the release of the chemical dopamine in certain key regions of the brain that is responsible for the rewarding effects of these drugs and that leads the user to wish to use them again. Others argue that this is too simplistic and that the significance of triggering dopamine release is that it may be "getting the brain's attention" to some significant stimulus (in this case, the psychotropic drug), and that this in turn may help determine the animal's motivation for seeking to repeat the experience. Furthermore, because alcohol and nicotine trigger dopamine release in the same way as THC, one could equally well argue that these, too, should be considered "gateway" drugs to cocaine, heroin, or amphetamines.

Adverse Effects on Fertility and the Unborn Child

For reviews of the information discussed in this section, see Maccarrone and Wenger (2005), Maccarrone et al. (2014), and Wang et al. (2006).

A paper published in the prestigious *New England Journal of Medicine* in 1974 sounded alarm bells (Kolodny et al., 1974). The authors reported that blood levels of the male hormone testosterone were severely depressed (average 56% of normal) in 20 young men who were regular marijuana users. In addition, some of the subjects were reported to have reduced sperm counts. These findings, of course, raised immediate concerns about the possibility that marijuana use might impair male sexual function or even lead to sterility and impotence. Numerous follow-up studies, however, either failed to repeat the original findings or found milder changes in testosterone levels or spermatogenesis (Zimmer and Morgan, 1997). There is little evidence for long-term infertility associated with marijuana use in humans, nor is there evidence of reduced fertility in countries in which heavy use of cannabis is endemic (Zimmer and Morgan, 1997).

Several studies have compared the babies born to women who had used marijuana during pregnancy with the babies of women who did not. Many of these studies failed to show any significant differences, but there is a consistent tendency toward a shorter gestation period and smaller birth weight in babies born to mothers who used marijuana. (Hayatbatbaksh et al., 2012). Similarly, a trend toward a higher incidence of birth abnormalities in the marijuana-exposed group in the same study may also not be considered statistically meaningful. If marijuana smoking does cause a reduction in birth weight, this is quite likely to be due to the presence of carbon monoxide in marijuana smoke. This gas binds tightly to the red pigment hemoglobin in the blood, making it less able to carry oxygen to the growing fetus. It is thought that the carbon monoxide in cigarette smoke is the most likely factor to account for the well-documented effect of tobacco smoking during pregnancy on reduced birth weight.

Several studies examined the development of children born to mothers who were exposed to marijuana during pregnancy to determine whether any abnormalities in physical or mental development could be detected. Although the results of the majority of these investigations were negative, the few instances in which subtle abnormalities could be detected in subsets of the IQ scale have been used as evidence that marijuana can impair children's cognitive development. In one of the largest studies of this kind, psychologist Peter Fried and colleagues examined a group of more than 100 children whose mothers were exposed to marijuana. In his Ottawa Prenatal Prospective Study, hundreds of different psychological tests were administered to the children, but very few differences were found between the marijuana-exposed and non-exposed groups (Fried, 1993). However, when the children were 6 years old, subtle deficits in frontal lobe executive functions were reported, involving visual memory, analysis, and integration, and these persisted when the children were examined again at ages 13–16 years (Fried et al., 2003). A further study of the cohort at ages 18–22 years using brain imaging and cognitive tests failed to detect deficits in working memory (Smith et al., 2006). The differences noted in the babies born to mothers who used marijuana were relatively minor by comparison with the consistent cognitive deficits observed in children of all ages born to mothers who had been heavy cigarette smokers during pregnancy; nevertheless, these findings were widely cited as one of the dangers of using marijuana (Fried, 1993).

Special Hazards of Smoked Marijuana

Traditionally, the use of cannabis in both Oriental and Western medicine involved taking the drug by mouth, but most of the current recreational and medical use of the drug in the West involves the inhalation of marijuana smoke by many users. Unfortunately, although smoking is a remarkably efficient means of delivering an accurately gauged dose of THC, it also carries with it special hazards. Although THC itself appears to be a relatively safe drug, the same may not be true for marijuana smoke.

The manner in which experienced users smoke marijuana tends to enhance the potential dangers of taking the drug by this route.

Marijuana smokers usually inhale more deeply than do tobacco smokers, and they tend to hold their breath in the belief that this increases the absorption of THC by the lungs. (In fact, the results of experimental studies in which both puff volume and breath-hold duration were systematically varied show that although inhaling more deeply does increase the amount of THC absorbed, holding the breath for more than a few seconds has rather little effect. The concept seems to be based more on cultural myths than on reality.) The results of these differences in smoking behavior are quite profound. Wu et al. (1988) compared the amounts of particulate matter (tar) and carbon monoxide absorbed in 15 volunteers who were regular tobacco and marijuana smokers. Results were compared after smoking a single filter-tipped tobacco cigarette or a marijuana cigarette of comparable size. Compared with smoking tobacco, smoking marijuana resulted in a fivefold greater absorption of carbon monoxide, and four or five times more tar was retained in the lungs.

It is possible that the use of higher potency marijuana may reduce unwanted tar deposition because smokers may inhale a smaller number of puffs (Herning et al., 1986; see Chapter 1). The conclusion seems to be that habitual marijuana smokers could reduce the health hazards of smoking by using marijuana with a high THC content. Other possibilities include the use of water pipes, filters, or vaporizers to reduce the tar content of marijuana smoke before it enters the lungs (see Chapter 2).

Because tobacco smoking is known to be the most important cause of chronic obstructive lung disease and lung cancer, it is reasonable to be concerned about the adverse effects of marijuana smoke on the lungs. There have been a number of attempts to address this question by exposing laboratory animals to marijuana smoke, but it is very difficult to extrapolate these findings to humans because it is difficult or impossible to imitate the human exposure to marijuana smoke in any animal model. The various studies that have been undertaken in human marijuana smokers seem far more relevant, although here the problem is confounded by the fact that many

marijuana smokers consume the drug with tobacco, making it difficult to disentangle the effects of the two agents. Professor Donald
Tashkin and colleagues at the University of California, Los Angeles
have been leaders of research in this field (for reviews, see Tashkin,
2005; Tashkin, 2013). In 1987, Tashkin reported the results of the
first large-scale study of 144 volunteers who were heavy smokers of
marijuana only (Tashkin et al., 1987). He compared these volunteers
with 135 people who smoked tobacco and marijuana, as well as
70 smokers of tobacco only and 97 non-smokers. Approximately
20% of both tobacco smokers and marijuana smokers reported the
symptoms of chronic bronchitis (chronic cough and phlegm production), even though the marijuana smokers consumed only three
or four joints a day versus more than 20 cigarettes for the tobacco
smokers. Ten years later, Tashkin et al. (1997) described an 8-year
follow-up study of the groups studied previously. They found that
lung function in the tobacco smokers had continued to get worse
over the period, particularly in the small airways, making them
more liable to develop chronic obstructive lung disease later in life.
However, no such decline was observed in the marijuana smokers,
suggesting that they may be less likely to develop such diseases as
emphysema because of their smoking. Similar conclusions were
reached in a study of 268 heavy marijuana smokers in Australia.
After smoking regularly for an average of 19 years, they had a lower
prevalence of asthma or emphysema compared to the general population. At the Kaiser Permanente health care group in California,
a careful comparison of 452 "daily" marijuana smokers who never
smoked tobacco with 450 non-smokers of either substance revealed
that the marijuana smokers had only a small increased risk of outpatient visits for respiratory illness (Relative risk = 1.19) (Polen
et al., 1993). In a large cross-section of US adults (from the National
Health and Nutrition Surveys, 2007–2010), cumulative lifetime use
of marijuana, up to 20 joint-years, was not associated with any adverse changes in lung function (Kemoker et al., 2015); this conclusion was also reached by examining a large cohort of US men and

women over 20 years in the Coronary Artery Risk Development in Young Adults (CARDIA project) (Auer et al., 2016). Tashkin (2013) reviewed the various longitudinal and cross-sectional studies and conclude that "habitual use of marijuana alone does not appear to lead to significant abnormalities in lung function."

A more sinister question is whether smoking marijuana might eventually lead to lung cancer. THC does not appear to be carcinogenic, but there is plenty of evidence that the tar derived from marijuana smoke is carcinogenic. Bacteria exposed to marijuana tar develop mutations in the standard Ames test for carcinogenicity, and hamster lung cells in tissue culture develop accelerated malignant transformations within 3–6 months of exposure to tobacco or marijuana smoke. Painting marijuana tar on the skin of mice also leads to premalignant lesions. But is there any evidence that this happens in the lungs of marijuana smokers? Is there any evidence for an increased risk of cancer in cannabis smokers? One of the few large-scale studies of the health consequences of marijuana smoking is that of Sidney et al. (1997). These authors studied a cohort of 65,171 men and women undergoing health checks at the Kaiser Permanente health care organization in California between 1979 and 1985. These subjects were then followed for an average of a further 10 years. Nearly 27,000 people admitted to being either current or former marijuana users (defined as ever having smoked more than six times). During the study period, 182 tobacco-related cancers were detected, of which 97 were lung malignancies. No effects of former or current marijuana use on the risk of any cancers were found. However, although this study involved large numbers, almost all the marijuana smokers were young (aged 15–39 years), and the follow-up period was relatively short. Such a study could not have been expected to detect any relationship between marijuana and lung cancer if the lag period were comparable to that seen with tobacco. Reviews of studies of the effects of marijuana smoking concluded that there was no evidence that marijuana smoking was linked to lung cancer (Zhang et al., 2014).

Huang et al. (2017) found no evidence that marijuana smoking was linked to lung cancer or a variety of other cancers. It is worth recalling that most of the studies which failed to find that marijuana smoking adversely affected lung function, and that it did not lead to the development of lung cancer, were performed with the relatively low-potency marijuana available 10–20 years ago; the impact of modern marijuana, containing several times more THC, might be very different. However, after reviewing the available data, Tashkin (2013) concluded that "findings from a limited number of well designed epidemiological studies do not suggest an increased risk for the development of either lung or upper airway cancer from light to moderate use."

Nevertheless, we should continue to be concerned about the possible link between marijuana smoking and lung cancer because it could take a very long time for such a relationship to become manifest. Cigarette smoking became common among men in the developed world during the first decades of the 20th century, but it was not until 30–40 years later that the first evidence of a link between tobacco smoking and lung cancer was obtained. Even though cigarette consumption has declined significantly in many developed countries, deaths from tobacco-related diseases will continue to rise for many years to come, particularly among women, for whom cigarette smoking was not common until the 1930s and 1940s. Such long lag periods between cause and effect are difficult to comprehend. The relationship between cigarette smoking and lung cancer is very complex. The increased risk of developing lung cancer depends far more strongly on the duration of cigarette smoking than on the number of cigarettes consumed each day. Thus, whereas smoking three times as many cigarettes a day does increase the lung cancer risk approximately 3-fold, smoking for 30 years as opposed to smoking for 15 years does not simply double the lung cancer risk but, rather, increases the risk by 20-fold, and smoking for 45 years as opposed to 15 years increases the lung cancer risk 100-fold (Peto, 1986; Peto et al., 1996).

The discovery of the link between cigarette smoking and lung cancer was one of the great achievements of medical research in the 20th century. The initial reports in 1950 from Britain and the United States were based on two very large case–control studies. Subsequently, a great deal more has been learned from follow-up studies in large groups of smokers and non-smokers. One such study involved asking all the doctors in Britain about their smoking habits. More than 40,000 doctors agreed to take part in a long-term study to determine what effect their smoking habits might have on their health. The study started in 1951, and Doll et al. (2005) described the results of a 50-year follow-up of this group. The results were alarming: Not only was the risk of dying from lung cancer increased in the cigarette smokers but also the risk of dying from 23 other causes, including cancers of the mouth, throat, larynx, pancreas, and bladder and such obstructive lung diseases as asthma and emphysema, was increased. The authors concluded that they had previously substantially underestimated the hazards of the long-term use of tobacco. The long-term follow-up data showed that approximately half of all regular cigarette smokers will eventually be killed by their habit.

Hamilton (2017) summarized one view of the hazards of cannabis use: "The greatest risk for most people who use cannabis continues to be exposure to tobacco. In the UK, we are uniquely wedded to the habit of combining tobacco with cannabis in a joint."

Summary

1. The safety profile of the active ingredient of cannabis, THC, is good. It has very low toxicity both in the short term and in the longer term. However, some of the acute effects of the drug, including the liability to cause unpleasant psychotic reactions in some and intoxication in others and to cause temporary impairments in skilled motor and cognitive functions, limit

the usefulness of THC as a medicine. There appears to be only a narrow "window" between the desired and the undesired effects.

2. Heavy use of marijuana can lead to schizophrenia-like symptoms and, in some cases, to marijuana psychosis that may persist for some days. There is an association between teenage use of marijuana and subsequent psychotic illness, but a causal link is still unproven.

3. Because of the cardiovascular effects of THC and its propensity to make schizophrenic symptoms worse, patients with cardiovascular disease or schizophrenia are not suitable subjects for cannabis-based medicines.

4. Whether there are persistent cognitive effects of marijuana even after prolonged abstinence, particularly deficits in verbal memory, remains contentious.

5. The adolescent brain appears more vulnerable to the adverse effects of cannabis on cognitive function.

6. The safety of smoked marijuana is questionable. It causes chronic bronchitis in a substantial proportion of regular users, but a link to lung cancer has not been observed.

The Institute of Medicine (1999) summarized the safety issues succinctly: "Marijuana is not a completely benign substance. It is a powerful drug with a variety of effects. However, the adverse effects of marijuana use are within the range of effects tolerated for other medications."

The Police Foundation review in 2000 also concluded, "By any of the major criteria of harm—mortality, morbidity, toxicity, addictiveness and relationship with crime—cannabis is less harmful than any of the other major illicit drugs, or than alcohol or tobacco."

Chapter 7

The Recreational Use of Cannabis

The use of cannabis as a recreational drug was almost unknown in the West until the 1950s and only became widespread during the 1960s. The exposure of large numbers of young American soldiers to cannabis during the Vietnam War was an important contributory factor (see, for example, the famous Vietnam War movies *Platoon* and *The Deer Hunter*). Just as Napoleon's army brought cannabis to Europe from its Egyptian campaign, the returning Vietnam War soldiers brought cannabis to the United States. The use of cannabis by young people on both sides of the Atlantic was closely linked to the protest and rebellion experienced by the 1960s generation. According to popular mythology, cannabis really started to enter the mass consciousness sometime in 1964 when the Beatles met Bob Dylan at an airport in America. He offered them a joint in the VIP lounge. Only Ringo Starr tried it then, but soon they were all enthusiastic users and role models. By the beginning of the new millennium, another generation has replaced the rebellious youth of 40 years ago—a new generation far less extravagant in its lifestyle. Members of this generation no longer believed the dire warnings issued by government about the dangers of marijuana. To this generation, marijuana was a part of its culture, no longer a gesture of rebellion. Many of the parents and grandparents of today's generation of cannabis users themselves belonged to the 1960s and 1970s group of marijuana smokers. The 1980s saw further attempts by governments to curb the use of drugs. The "War on Drugs" was waged by US Presidents Reagan and Bush beginning with the Anti-Drug Abuse Act of 1986, which introduced mandatory sentences for drug offenses. Despite this, 10 years later in 1996, California was the first state to approve the supply of medical marijuana. Cannabis is the most widely used illicit drug in the Western world. It ranks as the third most commonly used recreational drug after alcohol and tobacco. Whereas detailed information is available on the consumption of alcohol and tobacco—the health problems they cause and the consequent economic costs to society—such information is lacking in such precise detail for cannabis. Until recently, cannabis was used largely in an underground world of illegality. In most countries, according to the United Nations Conventions of 1961 and 1971 (which many countries signed), cannabis is considered as a dangerous narcotic—a Schedule I drug with no accepted medical uses. Possession of cannabis, cultivation of the cannabis plant, or trafficking are criminal offenses, some of which can carry severe penalties (United Nations Office on Drugs and Crime, 2017). This is still the official policy of many countries (including the United States and the United Kingdom), even if in practice cannabis-related offenses are treated leniently. It is not surprising that the users and suppliers of this illicit drug are not always willing to provide detailed information about it; nevertheless, millions of people are regular users.

Prevalence

There are some useful sources. The European Monitoring Centre for Drugs and Drug Abuse (EMCDDA, 2016b) provides annual data from 29 European countries. As many as half of the entire population aged 15–50 years in many Western countries admit to having used cannabis at least once. Consumption is highest in the younger age groups. In 2015, approximately one in five young European adults aged 18–25 years were current users of marijuana. The percentage of young adults who were current marijuana users in 2015 was stable compared with the percentages between 2011 and 2014.

However, the 2015 estimate was higher than the estimates from 2002 through 2010. Table 7.1 summarizes patterns of cannabis use among 15- to 34-year-olds in different European countries, using data from the latest available survey. Regular cannabis use is generally considerably lower than in the United States (less than half in most countries), with young people in the Netherlands, Italy, and Spain standing out as the most frequent consumers. In most European countries, approximately 5% of the adult population are current users (EMCDDA, 2016b). Spain has the highest cannabis consumption among young people (see Table 7.1), followed by Italy and the Czech Republic. In Spain, growing the plant on private property for personal use, and consumption by adults in a private space, is legal. In Denmark, cannabis is illegal, but it is freely available in the semi-autonomous area of "Christiana" in Copenhagen. In contrast, Sweden strictly bans cannabis use.

Patterns of consumption throughout the years have varied differently in various countries. In the United States, the National Survey on Drug Use and Health has produced a valuable annual report since 1972 (see Substance Abuse and Mental Health Services

TABLE 7.1 Percentage of European 15- to 34-Year-Olds Who Used Cannabis in the Past Month

Country	%	Country	%
Belgium	5.5	Italy	8.9
Czech Republic	8.5	Hungary	2.7
Denmark	6.4	Netherlands	8.3
Germany	6.3	Austria	3.4
Greece	1.5	Poland	4.8
Ireland	4.5	United Kingdom	6.1
Norway	3.0	Spain	12.0

Source: Data from the European Monitoring Centre for Drugs and Drug Abuse (2016b; available at http://www.emcdda.europa.eu/data/stats2016).

Administration (SAMHSA), 2016). Data are also available from the National Institute on Drug Abuse (NIDA) through its "Monitoring the Future" study (2016b); these statistics give a detailed picture of cannabis use among teenagers. Cannabis became very popular among young people in the United States during the 1970s, reaching a peak in 1979 when more than 60% of 12th grade students in US high schools (average age 18 years) admitted ever having used the drug, and 10% reported that they were daily users. There was then a marked drop in consumption during the 1980s as health fears grew. Consumption, however, increased again during the 1990s and now appears to have stabilized or to have declined somewhat in recent years (Bell et al., 1997; Compton et al., 2016; NIDA, 2016b). The NIDA data on American 18-year-olds in 2005 show that nearly one in five admitted to having used cannabis at least once during the past month, and 5% reported that they were daily users. Ten years later in 2015, the data were very similar, with 6% reporting daily use (Table 7.2).

Surveys in the United States reported that 13% of adults (>18 years)—one in eight people—used cannabis in 2016, and a

TABLE 7.2 Percentage of 12th Graders in the United States (Approximately 18 Years Old) Who Have Used Cannabis

Year	In Past Year (%)	In Past Month (%)
2009	20.6	5.2
2010	21,4	6.1
2011	22.6	6.6
2012	22.9	6.5
2013	22.7	6.5
2014	21.2	5.8
2015	21.3	6.0
2016	22.5	6.0

Source: Data from NIDA (2016b).

similar number was reported for 2014—significantly higher than the 10.3% reported in 2002 (Compton et al., 2016; Gallup, 2016; SAMHSA, 2016). Although this definition of "adult" is 18 year olds or older, it is worth noting that the minimum age for the purchase of legal cannabis is 21 years in states that have legalized cannabis.

In the United Kingdom, the British Crime Survey (2016) includes data on drug use. The data indicate that there was a statistically significant decline in cannabis use by young people (16- to 24-year-olds) during the period from 2000 to 2006, and there was a further decline between 2005–2006 and 2015–2016 (Table 7.3). The 2015–2016 British Crime Survey for the young adult population (aged 16–24 years) indicated that 7.7% had consumed cannabis during the past month, less than half the level recorded in 2001–2002.

It seems likely that only relatively small numbers of people on either side of the Atlantic use cannabis once a week or more, but the number using cannabis in the past year is estimated to be 11.3% of

TABLE 7.3 Percentage of 16- to 24-Year-Olds in Britain Who Have Used Cannabis

Year	In Past Year (%)	In Past Month (%)
2001–2002	27.3	17.6
2005–2006	21.4	13.0
2010–2011	17.1	9.0
2011–2012	15.7	9.2
2012–2013	13,5	n/a
2013–2014	15.1	n/a
2014–2015	16.4	8.5
2015–2016	15.8[a]	7.7[a]

[a]Decline in use between 2005–2006 and 2015–2016 is statistically significant.
n/a, not applicable.

Source: Data from British Crime Survey (2016).

adults in Britain (British Crime Survey, 2016; EMCDDA, 2017) and 5–10 million in the United States (see Table 7.3). The British government continues to maintain that cannabis is a harmful drug with no medical uses (i.e., Schedule I) (although approval of "cannabis oil" for limited medical use was given in July 2018).

How Is Cannabis Consumed and Where Does It Come From?

Traditionally, supplies of herbal marijuana or resin in the United States were from cannabis plants grown on farms in the southern United States and in northern Mexico. The US government's campaigns to eliminate these supplies (often by spraying the cannabis fields with the herbicide paraquat) led to imports from further afield. Colombia and certain Caribbean countries, notably Jamaica, became more important. In recent years, there has been a large increase in the consumption of home-grown cannabis—often using modern strains of plant yielding a high delta-9-tetrahydrocannabinol (THC) content and grown indoors. This has become a growth industry in the United States and Canada, with the expansion of the demand for legal medical marijuana and with outright legalization in some states increasing demand for recreational marijuana. Several crops can be harvested each year, making this a lucrative business. The most common form of the drug in the United States is marijuana "buds"(dried female flowering heads), nowadays frequently cultivated indoors, smoked or vaporized on its own. Rapid developments in cannabis products have made more potent preparations available recently (cannabis oil and concentrates; see Table 1.1). Most users prepare their own hand-rolled joints or "blunts," which are cigars emptied of their tobacco content and filled with marijuana, or they use vaporizers to deliver the THC from various cannabis products. The price of marijuana varies from state to state, with average prices for 1 ounce in the range $200–$350.

In the United Kingdom, most cannabis resin was traditionally imported from Morocco or other areas of North Africa, with smaller amounts from Pakistan, Afghanistan, Lebanon, and the Netherlands, but this has been overtaken by higher potency *skunk*, home-grown or imported illegally from the Netherlands (*nederwiet*). As in Canada and the United States, there has been a rapid increase in illegal cannabis farms in Britain. The costs of resin or herbal cannabis ranges from £150 to £300 per ounce. In most European countries, prices are similar. Cannabis is most commonly smoked with tobacco in hand-rolled joints or *spliffs*, although some herbal cannabis is smoked without tobacco, and resin may also be smoked in pipes or bongs or used in a vaporizer.

Countries or States That Have "Legalized" Cannabis

The Netherlands

Almost 50 years ago, in the 1970s, the Dutch government undertook the radical experiment of separating "soft" drugs such as cannabis from "hard" drugs such as heroin. By permitting the sale of cannabis, a separation was sought between soft and hard drugs. Soft drugs would be treated medically rather than criminally. This in turn led in the 1980s to the emergence of the "coffee shop," in which cannabis could be purchased and consumed on the premises or taken away. The term *gedogen* is at the heart of Dutch policy regarding cannabis; in a political context, gedogen means to tolerate or even permit an activity or behavior that is officially illegal. Enforcement of the law is considered to be a means to an end, not an end in itself, where rules are only enforced if the overall effects of doing so are considered to be positive. The coffee shop evolved as part of a deliberate policy to tolerate the sale and consumption of "soft" drugs (i.e., cannabis) compared to "hard" drugs such as heroin (Grether,

2017). The coffee shops were initially very popular, and more than 2,000 were opened in the Netherlands; the regulations were simple:

No hard drugs
Limit of 30 g herbal cannabis per day (later reduced to 5 g)
No hard advertising
No nuisance, loitering, noise
No underage sales (18-year-old limit)

The coffee shop model was popular and successful in limiting the number of heroin users and reducing heroin-related deaths, without leading to any significant rise of cannabis use in the Dutch population; the prevalence for cannabis use of approximately 8% remained similar to that of other European countries (see Table 7.1). In 1985, nearly 100% of methadone patients (treated for heroin addiction) were younger than age 40 years. In 2014, almost all of them were older than age 40 years. Thus, heroin use by people younger than age 40 years is practically non-existent, and the prevalence of heroin use is six times lower than in the United States. However, the policy was not always popular, especially because it drew "drug tourists" from far and wide to sample the Dutch cannabis experience. Whereas in tourist hot spots such as Amsterdam the tourist influx was beneficial to the economy, in others places it was less welcome. Particularly in towns such as Maastricht, on the border with neighboring states (Germany and Belgium), the influx of tourists coming to purchase legally available cannabis was becoming intolerable. The Dutch municipalities were given more freedom to regulate coffee shops, and more than half were closed down. In 2012, an attempt was made to limit sales of cannabis to Dutch nationals, who would be issued a *weitcart*, but this was not popular with those in large cities such as Amsterdam, although towns on the Dutch border took it up.

The Dutch cannabis experiment was flawed in not providing any legal sources of supply to the coffee shops, which had to rely on deliveries to the back door from criminal growers. As Amsterdam mayor van der Laan stated, "We have what we call a backdoor problem" (as cited in Grether, 2017).

Comparison of the pharmaceutical-quality medical products produced by Bedrocan for the cannabis pharmacies showed that the coffee shop material was often contaminated with pesticides and microorganisms (Hazecamp, 2006) or even sand or heavy metals such as lead (Knodt, 2017). Finally in 2017, the Dutch Parliament approved the legal cultivation of cannabis to supply the recreational market. If passed into law, the new rules would permit quality control of the cannabis sold in coffee shops, free up police time, and allow the government to levy taxes on cannabis supplies (Gilchrist, 2017).

Portugal

Portugal effectively legalized all drugs in 2001, and cannabis use is tolerated. Since then, the Portuguese example has been the source of wide debate of the advantages and disadvantages, regarded either as a success or a disastrous failure (Hughes and Stevens, 2010). Although not strictly a model for cannabis, because Portugal effectively decriminalized all drugs, it is notable that there has not been any significant increase in the prevalence of cannabis use, which is slightly more than 5% for young people and adults (Transform, 2014). Predictions that Portugal would be flooded with drug tourists have not materialized.

Uruguay

Uruguay was the first country in the world to fully legalize the production and sale of marijuana. The law was passed in December 2013 after a decade-long grassroots movement headed by mostly middle-class consumers managed to convince the government it was safer to legally sell cannabis rather than to allow drug dealers to run the market. The system grants licenses to private producers for large-scale cannabis farming and regulates its distribution at a controlled price of approximately $1 per gram through pharmacies to registered consumers. Private individuals are also allowed up to six plants at home. Larger amounts can be grown at "cannabis clubs," where individuals band together to produce marijuana in greater quantities, as long as it is not for sale.

However, the new law was not an immediate success. By July 2014, when legal sales started, only 50 of 1,200 pharmacies had registered to sell cannabis. Many pharmacies were concerned about robberies, increasing costs and paperwork, and disclosing their activities to customers who did not support legal marijuana. Although the price is highly accessible, only $1.30 per gram compared with $3 on the street, consumers must first register with the government. They are then required to identify themselves with a digital thumb scan to withdraw their weekly maximum of 10 g. The registry was opened at the beginning of May 2014. To date, approximately 3,500 people (out of Uruguay's population of 3.4 million) have signed up to buy marijuana at pharmacies. In addition, since 2014, approximately 6,700 have signed up as home growers, and 57 cannabis clubs have been set up, according to the government's Cannabis Regulation and Control Institute. Because of the privacy issues involved, the government has taken important precautions to prevent the registry from falling into the wrong hands. Some pharmacists nevertheless believe that they may become targets of violence as their pharmacies, with fixed-price cannabis, are taking away the profit margins from street corner dealers. It remains to be seen if this adventurous move will be a success or a disastrous failure (Goni, 2017).

The United States

In the US election in November 2016, Massachusetts, Nevada, and California joined Colorado, Washington, Oregon, Alaska, and the District of Columbia by voting for initiatives that make it legal for adults to consume cannabis (Mic, 2017). The margin in Maine was very close (50.3% for vs. 49.7% against) and required a recount; Governor Le Page threatened to overrule the decision and implementation was delayed until 2018. If the initiatives are all implemented, more than half of adult Americans will have access to legal marijuana. This outcome is by no means certain, however, because federal law, which states that cannabis is a dangerous drug with no medical applications, takes precedent over state laws. As of 2017, President Trump and Attorney General Jeff Sessions were adamantly opposed to legalizing cannabis. On March 16, 2017, Sessions stated,

> I reject the idea that America will be a better place if marijuana is sold in every corner store. And I am astonished to hear people suggest that we can solve our heroin crisis by legalizing marijuana—so people can trade one life-wrecking dependency for another that's only slightly less awful. Our nation needs to say clearly once again that using drugs will destroy your life.

There is little doubt that the Justice Department could halt or seriously interfere with the moves to decriminalize cannabis, despite opinion polls showing that 60% of Americans support relaxing the law (Swift, 2016), and many voters used their democratic right to say so in elections. Nevertheless, the example of Colorado is examined in detail here because it was one of the first states to legalize, and more than 5 years of experience may indicate what effect the legalization of cannabis may have elsewhere in the future.

Colorado approved the "legalization" of cannabis in 2012, and implementation started in January 2013. The regulations and laws governing access to legal marijuana in Colorado have been refined continuously. Anyone who is 21 years old or older has a constitutional right to possess and consume marijuana in Colorado. A government-issued identification is required to prove one's age, so a driver's license or passport is sufficient. One does not need to be a Colorado resident to possess cannabis, and there is no registration system. Previously, tourists in Colorado were restricted to purchasing 7 g or less, whereas Colorado residents could purchase up to 28 g. This law changed in June 2016, and now both tourists and residents can purchase 28 g in a single transaction. There are more than 500 "retail outlets" where cannabis and "edibles" can be purchased (see Chapter 1). Schedules of the THC content of equivalent amounts of edibles or concentrates with herbal marijuana have been published. The consumption of cannabis is not permitted openly or publicly. In general, there are no coffee shops or marijuana bars as one might find in Amsterdam, nor can marijuana be consumed in bars or concert venues. The exceptions are the "smoking lounges" of private members clubs, where one may purchase a membership for 1 day. Many users circumvent these rules by the increasingly popular use of personal vaporizers, which leave no odor and can thus be used anywhere without detection. Other restrictions on marijuana use include a "driving under the influence" law, which sets a legal limit of 5 ng THC per milliliter of blood—with serious penalties for infringement, including loss of driver's license. Possession of marijuana is not allowed on any federal land (national parks, national forests, etc.), with mandatory sentences for infringement. Colorado residents are permitted to cultivate up to six plants, three of which can be in the flowering stage, kept in a locked space. These regulations, which have evolved over the 5 years since legalization was approved, seem to provide a sensible model that other states may wish to follow.

In Colorado, legalization has resulted in other changes in the way society operates. These are what may be described as the "exuberant reactions to novelty." For example, in Colorado one may check into a marijuana-friendly hotel in which the concierge can offer advice on cannabis tours, and guests are able to rent a vaporizer; they may also wish to indulge in a massage with cannabis-infused oil. One can also register for a course in cannabis cookery or advanced cannabis cookery. In addition, people can use catering services to provide their guests with marijuana-containing meals.

In Colorado, there are more than 500 retail stores in which cannabis and related products can be purchased. They offer a bewildering variety of products. For example, the Botanico–Adult Use store in Denver offers 21 strains of *Cannabis indica*, with such exotic names as Grape Ape and Kurple Fantasy; 6 strains of *Cannabis sativa* (e.g., Golden Goat and Blue Dream); and 7 other hybrid strains. Elsewhere, extravagant claims are made for various strains: The popular Purple Kush is described as "soothing and relaxing" and "great for sleep, pain, stress and nausea"; Sour Diesel is said to provide "uplifting and visual experience" and to be "great for mood and appetite"; and Dutch Crunch is claimed to have "mood enhancing qualities" and to encourage "clear-thinking and productivity." In the Botanico–Adult Use store, the THC content listed for each product is in the range 15–21%, with no data on cannabidiol—probably vanishingly small. The different varieties of "concentrates" with even higher THC content (see Chapter 1) range from cartridges for vaporizers to wax, oil, or solid materials. The store also sells 29 "edibles" containing THC, from lollipops to mints that are sucked, chews, a variety of truffles and other sweets, and many flavored drinks. (A major criticism of the edibles market is that children may access sweet drinks or sweets and accidentally overdose with cannabis.) One can also purchase "topicals" in the form of transdermal patches, oils, and balm. In addition, vaporizers and cartridges containing 250–500 mg THC are available. At first sight, this may seem an excessive range of products, but no more so than any food supermarket, offering thousands of different foodstuffs.

A widely believed statistic cited in a report by marijuana policy researcher Jon Gettman (2009) is that the cultivation of marijuana is the largest cash crop in the United States, valued at $35.8 billion in 2006—larger than corn ($23.3 billion) or soybean ($17.3 billion). This estimate has been widely quoted by both those who favor legalization and those who oppose it. However, the assumptions made by Gettman about both the extent of marijuana production and farm prices may have been wildly exaggerated; the actual cash value of the marijuana crop in the United States is likely to be in the range $2.7–$4.3 billion (Caulkins et al, 2016). Even this estimate places marijuana in the top 15 cash crops in the United States, and there is little doubt that production will rapidly escalate as legalization proceeds. Colorado found that retail sales of medical marijuana remained almost constant, but sales of recreational marijuana grew rapidly in the first 3 years since legalization: $699 million in 2014, $996 million in 2015, and $1.3 billion in 2016. Sales are predicted to continue climbing for some time, possibly overtaking tobacco by 2020, and so far the state of Colorado has received almost $200 million in marijuana tax revenues—a figure that other states are keen to emulate.

A detailed review by the Marijuana Policy Group (2016) estimated that in 2015, the economic impact of marijuana legalization in Colorado was $2.39 billion and that the marijuana industry had generated 18,000 jobs. The cultivation, distribution, and sale of legal marijuana have become a new industry. The commercial business of marijuana has led to the formation of new companies, some of which are listed on the New York Stock Exchange (Wealth Daily, 2017). Large new growing and processing facilities have been announced in Washington state (Williams, 2015), California (Lindsey, 2016), and Massachusetts (Livni, 2016)—the latter covering 1 million square feet, which will make it the largest in the country. These figures suggest that the nationwide legalization of marijuana could generate much-needed jobs and tax revenue. However, it is worth noting that the data from Colorado are inflated by the development of a flourishing cannabis tourist industry, the size of which is unclear. It must also be remembered that the legalization process in US states is still at variance with federal law and may be halted or slowed by government.

Canada

In April 2017, Canadian Prime Minister Trudeau announced that cannabis would be legalized in Canada, effective July 1, 2018. This makes Canada the second country in the world to do so. The impact of having marijuana legalized on such a large scale and so close to the United States may affect the future of the US marijuana industry. The lessons that Canada will learn while setting up the market— how rules vary among provinces that share borders with each other or with the United States—will be ones from which the United States can learn, just as states such as Oregon and Washington have had the benefit of learning from Colorado's growing pains with regard to regulatory issues. Canada's recreational marijuana task force consulted with Washington and Colorado to prepare for legalization. Investors are keen to invest in Canadian cannabis companies, wary of spending their money in the United States as long as the federal government views marijuana as a drug on the same legal footing as heroin (Steinmetz, 2017).

The task force reported that the following rules should be applied to cannabis regarding growth, ownership, and possession:

a. The national age requirement to purchase cannabis will be set at 18, but provinces will be able to increase the age limit as they see fit.
b. In addition to the age limit, provinces will control the price, as well as the manner in which cannabis is bought and sold.
c. The federal government will be in charge of licensing producers, as well as regulating the supply of cannabis across the country.
d. A limit of four plants per household will be established for those who wish to grow their own recreational cannabis.

The federal government will be in charge of licensing producers, as well as regulating the supply of cannabis throughout the country. The government should also apply comprehensive restrictions to the advertising and promotion of cannabis and related merchandise by any means, including sponsorship, endorsements, and branding, similar to the restrictions on promotion of tobacco product under the Tobacco Act. The government should expect edibles to have a broad appeal. Cannabis products such as brownies, cookies, and high-end chocolates are attractive to novice users and those who do not want to smoke or inhale. It is important to ensure that cannabis edibles can be clearly distinguished. It can be a challenge to differentiate between cannabis edibles and cannabis-free products, leading to a risk that individuals, including children, will inadvertently consume them (Veksler, 2017). Beginning July 2018, adults who are age 18 years or older will be able to

possess up to 30 grams of legal dried cannabis or equivalent in non-dried form

share up to 30 grams of legal cannabis with other adults

purchase dried or fresh cannabis and cannabis oil from a provincially-licensed retailer (In those provinces that have not yet or choose not to put in place a regulated retail framework, individuals would be able purchase cannabis online from a federally-licensed producer.)

grow up to 4 cannabis plants, up to a maximum height of 100 cm, per residence for personal use from licensed seed or seedlings

make cannabis products, such as food and drinks, at home provided that organic solvents are not used

other products, such as edibles, would be made available for purchase once appropriate rules for their production and sale are developed. (Canadian Government, 2017)

The federal, provincial, and territorial governments will share responsibility for overseeing the new system. The federal government's responsibilities will be to:

set strict requirements for producers who grow and manufacture cannabis

set industry-wide rules and standards, including

- the types of cannabis products that will be allowed for sale
- packaging and labelling requirements for products
- standardized serving sizes and potency

- prohibiting the use of certain ingredients
- good production practices
- tracking of cannabis from seed to sale to prevent diversion to the illicit market
- restrictions on promotional activities (Canadian Government, 2017)

In summary, Canada put cannabis legalization in place nationwide 1st July 2018, giving the provinces considerable latitude to set their own local rules. The regulations controlling cannabis will be similar to those in place for tobacco. Both supply and consumption will be tightly controlled. The world awaits the outcome of this major change. Unlike the United States, the legalization scheme originates from the Canadian federal government, so there should be less ambiguity about the legal status of the process.

Snapshots of Cannabis Use Outside Europe

Cannabis has been used for hundreds of years in different countries and cultures both for recreational and medicinal uses and as an integral part of religious rites (for reviews, see Rubin, 1975; Robinson, 1996; Booth, 2003). An understanding of this may help us to place the modern vogue for cannabis use in the Western world into a broader context. Many modern users of cannabis speak of their feelings of spirituality and "oneness with God" when intoxicated. Cannabis is used in many religions as a sacrament, from the "dagga cults" in Africa to the Ethiopian Copts, Hindus, Zoroastrians, Rastas, Buddhists, Taoists, and Sufis. Unlike drinking alcohol, the use of cannabis is not expressly forbidden in the Koran, and in some Moslem countries cannabis is widely used.

India and Pakistan

The report of the Indian Hemp Commission (1894) gave a detailed account of the use of cannabis in the Indian subcontinent more than

100 years ago, and Chopra and Chopra (1957) described a situation that seemed to have changed little more than half a century later. However, cannabis was made illegal in 1985, except in government-authorized shops that sell *bhang*, which can refer either to the herbal cannabis balls or to the drink made by mixing it in milk. All production and selling of bhang has to be through government-authorized shops. Bhang has a relatively low THC content, and it is usually considered a sleep inducer rather than a recreational drug. Used during the observance of certain Hindu rituals, government-owned shops in holy cities such as Varanasi sell cannabis in the form of bhang. Despite the high prevalent usage, the law makes it illegal to possess any form of cannabis. However, this law is rarely enforced and is treated as a low priority throughout India. Furthermore, large tracts of cannabis grow unchecked in the wild in many areas of northern and southern India, including states such as West Bengal, Tripura, Andhra Pradesh, Karnataka, Kerala, and Tamilnadu. Also, many states, such as West Bengal and Tripura, and the North East have their own laws allowing cannabis, locally known as ganja. Bhang is often used to make a beverage called *thandai*—many variants of this drink exist. Bhang may be mixed with many other ingredients, including milk, almonds, melon, poppy seeds, aniseed, cardamons, musk, and essence of rose. Sweetmeats containing bhang and even ice cream containing the powdered leaves may also be used. Whereas alcohol is generally looked down upon in Hindu society, high-caste Hindus are allowed bhang at religious ceremonials and also employ it as an intoxicant at marriage ceremonies and family festivals. Bhang is used in the Hindu religion in particular to celebrate the last day of the Durga Puja, and offerings are made to the god Shiva in Hindu temples. Itinerant Hindu ascetics also use bhang. Laborers in India use cannabis traditionally in much the same way as beer is used in the United States. A few pulls at a ganja pipe or a glass of bhang at the end of the day relieves fatigue and provides them with a sense of well-being to enable them to bear more cheerfully the strain and monotony of their daily routines.

Bhang is probably equivalent to low-grade marijuana in its THC content, and the watery infusions that are drunk probably contain rather little active drug, although milk (which contains fats) would be a more effective means of extracting THC. Intoxication after taking bhang is uncommon, and the Indian Hemp Drugs Commission's conclusion in 1894 that the moderate use of hemp drugs caused no appreciable physical, mental, or moral injury is probably still correct.

In Pakistan, although cannabis is illegal, the export of cannabis products is considerable. Growers in the Tirah Valley in Pakistan's tribal belt export some of the world's most sought-after hashish (Craig, 2015).

Nepal and Tibet

The advent of the hippie era and the migration of young Westerners to the Himalayas in the 1960s in search of cannabis and spiritual enlightenment led to some remarkable changes in local attitudes to cannabis in these cultures. In Nepal, cannabis was legally available and was traditionally used by Hindu yogis as an aid to meditation, and male devotees used it as a symbol of fellowship in their communal consumption of the drug. Older people also used it to while away the time when they were too old to work in the fields. The advent of the hippie era and an influx of Westerners, however, brought about increased cultivation of cannabis, inflated prices, and a change in attitude of young, middle-class Nepalese to the extent that smoking cannabis came to be regarded as a novel, acceptable, and pleasurable mark of sophistication. This in turn led a panic-stricken government in Nepal to introduce harsh new laws during the 1970s in an attempt to suppress the use of the drug (Rubin, 1975), although an illegal trade continues.

In Tibet, cannabis plays a significant role in some Buddhist ceremonies. According to Indian tradition and writings, Siddhartha used and ate nothing but hemp and its seeds for 6 years prior to announcing his truths and becoming the Buddha in the 5th century BC.

Southeast Asia

Cannabis is common in Cambodia, Thailand, Laos, and Vietnam—many Americans were introduced for the first time to the drug during military service in Vietnam in the 1960s and 1970s. The laws prohibiting the legal use of cannabis are strong but are not rigorously enforced. The herbal material is often used in the local cuisine to impart an agreeable flavor and mild euphoriant quality. Medically, cannabis is recognized as a pain reliever and is used in the treatment of cholera, malaria, dysentery, asthma, and convulsions. Cannabis is considered to be a source of social well-being, to be shared with friends, and is also used to ease difficult work tasks (Rubin, 1975). Japan, South Korea, and China have strict laws prohibiting cannabis, but lobby groups in Japan are pressing for reform. In China, although cannabis is illegal, there is a long history of expertise with herbal medicines. Chinese companies own more than half of the 600 patents relevant to medical cannabis filed with the World Intellectual Property Office. The country is well positioned to dominate the global medical cannabis market (Johnston, 2014).

Africa

The use of cannabis both for pleasure and for religious purposes is common throughout most of Africa, where it predates the arrival of Europeans. Known commonly as *dagga*, cannabis is a sacrament and a medicine to the Pygmies, Zulus, and Hottentots. Its use in religious ceremonies in Ethiopia is ancient, and it was taken up and used as a sacrament there by the early Coptic Christian church.

In Morocco, cannabis, known as *kif*, is traditionally served as a stimulant and as a means of relieving the pressures of daily life among the tribal groups living in the Rif Mountains. The growing of cannabis in this northern region of the country, which was previously the poorest agricultural area, has become an important agricultural export industry for this region. With modern strains of plant and improved "extractors" to harvest the resin bracts in the preparation of high-potency resin, growers attempt to remain internationally competitive.

Caribbean

Jamaica has become an important cultivation center for cannabis. The drug, known as ganja, was brought there by laborers from India in the mid-19th century and spread to the Black working-class community, where its use has become widespread. It is also exported to the United States. Ganja smoking is so prevalent among working-class males that the non-smoker is regarded as a deviant. The occasion of first smoking attains the cultural significance of an initiation rite and ideally should be accompanied by the "ganja vision." Jamaica is also home of a 20th-century religion known as Rastafarianism, founded by Marcus Garvey in the 1930s, in which cannabis plays a key role. Members of this religion, known as Rastas, accept some parts of the Bible but believe that the Ethiopian Emperor Haile Selassie was a living God and represented "Jesus for the Black race." Ethiopia is thought of as the ancient place of origin of Black people, and an eventual return to Ethiopia would be their equivalent of nirvana. The ritual smoking of cannabis forms a key part of the Rastafarian religion; it is thought to cleanse both body and mind, preparing the user for prayer and meditation. Rastas, with their characteristic dreadlocks and their dedication to cannabis, have permeated many aspects of modern culture, especially in the field of pop music. One of the most famous was the musician Bob Marley, who died in 1981.

South America

Cannabis smoking is common in many Latin American countries. In some cases, as in Brazil, it was brought there by African slaves and spread among the working people as "the opium of the poor." In Mexico and Colombia, the cultivation of cannabis for export has become an important cash crop, and along with this has come widespread use of the drug. Whereas marijuana smoking in Colombia was formerly regarded as socially undesirable, it has become acceptable in many circles. Mexican outdoor growers provide a major export of illegal marijuana to the United States, although public opinion in Mexico is veering toward legalization. A 2017 survey published in the *International Journal of Drug Policy* places support for legalization in Mexico at 41%, compared to 29% in 2016. Mexico's Senate passed a medical marijuana bill in December 2016, although the move still needs the lower house's approval. Mexico has been the home of a number of religious sects that use cannabis as a sacrament. For example, members of a small community near the Gulf of Mexico use marijuana, which they call "la santa rosa," in their religious ceremonies. The dried herbal cannabis rests on the divine altar wrapped in small bundles of paper, along with artifacts of ancient local gods and images of Catholic saints. The men and women priests of the church chew small quantities of the herb, and it gives them inspiration to preach to the congregation. The French anthropologist Louis Livet (1920) described a remarkable communal marijuana ritual among a sect of native Indians in Mexico. Participants were seated in a circle, and each in turn took a puff at a large marijuana cigar, which he passed to his neighbor. The atmosphere at such meetings was joyful and filled with ritual chanting and convivial warmth. Each of those attending took a total of 13 puffs, and at the end consequently found themselves in a state of hallucinatory excitement and intoxication. At the center of the circle was placed a sacred animal, an iguana. The animal, attracted by the smell of the marijuana smoke, rotated 13 times, turning its head toward the cigar with its mouth open, inhaling the smoke. The animal was thought to represent the sacred incarnation of a god presiding over the ceremony, and when the iguana became intoxicated and fell down, the participants knew that it was time to stop passing the cigar. The reptile served a function akin to that of the pit canary in 19th-century coal mines.

Several South American countries have opted to take steps toward legalization. Most recently, Argentina's legislature approved marijuana for medicinal use in 2017. Chile and Colombia took the step of passing legislation allowing the use of medical marijuana in recent years, and Colombia also plans to offer a crop substitution program for farmers of illegal coca crops to cultivate marijuana legally as part of its peace process with the Revolutionary Armed Forces of Colombia, better known as the FARC. This changing mindset shows that some countries are seeking alternatives to police and military intervention when it comes to marijuana (Gonzalez, 2017).

Russia

Russia has the one of the world's most intolerant drug policies, with no distinction between soft and hard drugs and lengthy prison sentences for dealing and trafficking. Called a "dangerous gateway

TABLE 7.4 Most Common Positive Benefits Reported by 2,794 British Cannabis Users

Effectss	*% Reporting*
Relaxation/relief from stress	25.6
Insight/personal development	8.7
Antidepressant/happy	4.9
Cognitive benefit	2.9
Creativity	2.3
Sociability	2.0
Health Effects	
Pain relief	6.1
Respiratory benefit	2.4
Improved sleep	1.6
Total reporting positive effects	57.8

TABLE 7.5 Adverse Effects Attributed
to Cannabis Experienced Regularly by British Users
in 1999 Independent Drug Monitoring Unit Survey

Effect	% Reporting
Apathy	20.9
Balance	8.3
Paranoia	6.4
Impaired memory	3.9
Anxiety/panic	3.8
Chest problems	3.5
Withdrawal	1.5
Total reporting problems	21.0

drug" by Russia's Federal Drug Control Service, the authorities made it clear that there are no plans to legalize marijuana. Russia has mobilized the BRICS nations (Brazil, Russia, India, China, and South Africa) against marijuana. The members are all developing or newly industrialized countries with a significant influence on regional and global affairs (*Marijuana News*, 2016).

Conclusion

It seems likely that cannabis will be available legally in many areas of the world in the near future (see Chapter 8). It is worth noting that warnings of the harmful effects of cannabis tend to be emphasized, but more than half of British cannabis users reported positive effects (Table 7.4), whereas only one in five commented on adverse effects (Table 7.5).

Chapter 8

Where Are We and Where Are We Going?

So much has happened in the field of cannabis research since the second edition of this book was published in 2008 that it is perhaps appropriate to look initially at some of the advantages and disadvantages of increased medical and recreational use of marijuana. Much of this debate has originated in the United States, where both medical and recreational marijuana have been approved by a majority of the population.

The Case for Medical Marijuana, and Some Cautions

Perhaps the most powerful reason in support of medical marijuana lies in the legitimate conditions for which "hard" data are available, while recognizing that a far wider range of medical uses may be treated, for which the supporting data are weak but marijuana may act as positive placebo. The safety of marijuana is another positive feature: Physicians are unlikely to harm their patients by prescribing marijuana, and there may be real benefits. On the other hand, the influential *Journal of the American Medical Association* published an editorial in 2015, directed to American physicians, titled "Medical Marijuana: Is the Cart Before the Horse?" with the following warnings (D'Souza and Radhakrishnan, 2015):

1. States in which medical marijuana is legal list a number of approved medical indications, for most of which the evidence is of poor quality.
2. The risk of marijuana use causing neuropsychological deficits is increased in young people.
3. Unlike most medicines, which contain a single active substance, herbal marijuana contains many different cannabinoids and other natural products of unknown properties.
4. The synthetic cannabinoids dronabinol and nabilone are already approved medicines available for physicians to prescribe.
5. The prolonged use of marijuana carries a risk of addiction and a lesser risk of the development of psychosis.
6. The interaction of marijuana with other medicines has not been systematically studied.
7. Marijuana activates endocannabinoid mechanisms that are key to normal neural development.
8. The nature of the effects of marijuana is unclear. What physiological mechanisms are involved?

In summary, why are physicians asked to prescribe a medicine that has not gone through the rigorous safety testing and clinical trials normally expected by the US Food and Drug Administration (FDA)? It is true that medical marijuana is not an FDA-approvable medicine; nevertheless, it is here to stay, and physicians and others must find the best way to regulate it.

The Case for Allowing Cannabis for Recreational Use

The arguments in favor of legalizing the recreational use of cannabis are more complex. Where there is no medical need, we must be sure that the policy is not damaging users and that young people can be protected. The influential newspaper, *The New York Times*, published a series of editorials in 2015 under the title "Six Powerful Reasons to Legalize Marijuana." They are as follows:

1. Prohibition has enormous costs associated with it, in terms of police activity, prisons, and so on. A total of 643,122 marijuana-related arrests were made in 2015 (down from a peak of 872,721 in 2007).
2. The benefits of the criminalization of marijuana are negligible or non-existent.
3. Prohibition has racist implications: Mexicans or African Americans are more likely to be arrested for marijuana-related offenses in the United States.
4. Marijuana has legitimate medical uses, although federal law describes it as a Schedule I drug with no medical use.
5. Evidence from Colorado and other states that have legalized marijuana, and from European countries in which it is legal, shows that legalization does not lead to increased marijuana use.
6. Marijuana is less harmful than alcohol or tobacco, which are linked to hundreds of thousands of deaths annually.

Future Scenarios

If marijuana continues to remain legal in the states in which it is currently legal and legalization proceeds in other states whose voters approved it in the 2016 election, this would be a major political and social change—equivalent to the shift in social attitudes toward smoking tobacco. Not knowing which way this debate will turn out, perhaps a summary of the positive and negative outcomes is useful.

Positive

One view of the future is positive: The era of cannabis prohibition is coming to an end. This is well reviewed by Alyson Martin and Nushin

Rashidian in their entertaining monograph, *A New Leaf: The End of Cannabis Prohibition*" (2017). Among other things, the authors provide insights into the political movements and discussions underlying the debate on the relaxation of cannabis laws in the United States. The end of cannabis prohibition would have obvious benefits in the United States, including the additional state revenue yielded by cannabis sales tax and other related taxes. For example, the state of Colorado has generated $200 million in tax revenue during the first 4 years since legalization. Estimates suggest that cannabis sales tax could raise $58–$100 million initially in California. In addition, the costs of maintaining prohibition nationally could be greatly reduced. The budget of the Office of National Drug Control Policy includes $89.6 billion annually for domestic enforcement of drug laws, mainly cannabis arrests. Such arrests carry a lifelong record of drug misdemeanor charges, which no matter how trivial can result in reduced income, housing, child custody, and student financial aid. The US budget also includes $5 billion to fight the drug war internationally, mainly in Mexico. The current policy has distinct racial connotations because four times as many blacks as whites are arrested for possession.

Released from prohibition, the supplies of cannabis would be removed from the current largely uncontrolled criminal sources, and a series of national standards could be set for the cultivation and quality of cannabis on sale, with regular inspections and tests administered by a government agency (as in Canada). The price of "legal" cannabis could be set sufficiently low to make criminal supplies of illegal cannabis uncompetitive. The availability of high-quality cannabis would also set researchers free from legal constraints to work on the major scientific questions still unanswered. For the first time, researchers could study the positive effects of the drug. From such research, new medicines, based on our new scientific understanding, will eventually emerge, and the pharmaceutical industry will have a renewed interest in the market opportunities opened by research.

The new era of legalized cannabis will carry responsibilities, particularly in the regulation of laws on the availability and use of cannabis, aimed at protecting young people from potential harm. There may be a case for reducing the minimum age to 18 years, rather than the current 21 years in some states, because illegal cannabis use by 18-year-old high school students is already high. The regulation of cannabis use will no doubt pose a series of new issues concerned with questions regarding advertising, intrastate transport, the use of cannabis in public places, and so on. Five years into the era of legalization, Colorado is still finding that it needs new laws to regulate the cannabis market. There is little doubt that effective regulatory regimes can be developed. Society has done this successfully for tobacco and alcohol: Their use in public places has been restricted, advertising has been limited, and the use of either drug by young people is on the wane.

If cannabis becomes a legal product, there will be commercial opportunities for its large-scale production and packaging for profit by private or public companies, in addition to opportunities to profit from the many other needs of a cannabis culture. Such commercial activity is already being developed; it is hampered by the laws that prevent banks and insurance companies from handling the accounts of companies or groups of individuals involved in the cannabis trade. The repeal of such restrictive laws should place the cannabis industry in a more normal commercial position compared to its current cash-only commerce. The experience of Colorado has shown that the cannabis industry can add many much-needed new jobs and new revenues to the economy.

The legalization of marijuana in Canada is an important change in policy for a country with close ties to the United States. If successful, it could lead to other US states approving legalization.

In Europe, there is an increasing social movement to relax restrictions on cannabis, with some countries already having legalized it(the Netherlands and Portugal).

Negative

Although a majority of those polled in the United States are in favor of relaxing cannabis prohibition (60% for vs. 40% against), this does not represent a uniform geographical picture. Some states continue to favor prohibition, and even in those with a majority in favor of ending prohibition, there are some who are firmly opposed. In Maine, for example, Governor Paul LePage was firmly opposed to the medical use of marijuana and he persuaded the legislature to delay approval of any relaxation. Legislators in a number of states have pushed forward measures to delay the enactment of voter-initiated marijuana laws (Kindland, 2017).

In the current era of President Trump and Attorney General Jeff Sessions, the case for an end to prohibition in the United States seems remote. Sessions and Trump have declared themselves strongly opposed to the legalization of recreational marijuana, although possibly allowing medical uses. Attorney General Session has declared himself adamantly opposed to legal marijuana and has conflated marijuana use with the current opiate overdose crisis in the country. He is in favor of reactivating the "War on Drugs" in the United States and has said he will do everything possible to disrupt the marijuana industry. Members of the Justice Department's Task Force on Crime Reduction and Public Safety have been ordered to "undertake a review of existing policies" regarding federal marijuana law enforcement, among other things. In its June 2017 newsletter, the National Organization for Reform of Marijuana Laws listed 11 ways by which Trump and Sessions could disrupt the legalization of marijuana, including the use of an old piece of legislation, the Racketeer Influenced and Corrupt Organizations Act (RICO). A recent court victory by two Colorado landowners who complained that the smell from a nearby outdoor marijuana farm made horse riding on their property less pleasant advanced a strategy, using this statute, that anti-marijuana forces hope will help them destroy the marijuana industry in Colorado and throughout the country.

In addition, in a May 2017 letter to congressional leaders, Sessions asked them to roll back federal protections for the drug that have been in place since 2014. Known as the Rohrabacher–Farr amendment, these protections prevent the Justice Department from using federal funds to block states from crafting their own medical marijuana regulations. In his letter, Sessions complained that the amendment was keeping his department from enforcing other federal laws—namely the Controlled Substances Act—citing the country's "historic drug epidemic and potentially long-term uptick in violent crime" as justification.

There is little doubt that the US federal government can use its overriding power to achieve its ends, although this policy would be at variance with the popular vote and is hardly likely to go unchallenged. Nevertheless, as Sessions said, "Good people don't smoke marijuana."

Legalizing cannabis does not necessarily mean that people can purchase legal cannabis. In 2013, Uruguay was hailed as the first nation to legalize marijuana. However, the policy has not proved very effective. The government wanted all users, even home growers, to register, and many were frightened to do this in case of violent reprisals by members of the flourishing illegal cannabis market. The government persuaded only a handful of pharmacies to provide legal cannabis at a low price—with no nationally organized source of supply. The illegal industry demonstrated how powerful vested interests could be in blocking a government policy.

A similar situation may exist in the Western world. Big Pharma has shown little interest in marijuana, although it is keen to develop synthetic drugs that rely on knowledge of the science of marijuana. Big Pharma, the tobacco companies, and the alcohol industry must view with some concern the advent of a competing product, which will inevitably erode some part of their market. and these industries have huge lobbying power not available to the marijuana industry.

Despite increasing support for marijuana legalization, the US Federal Bureau of Investigation reported 750,000 marijuana arrests (mainly for possession) in 2012, 701,000 in 2014, and 643,122 in

2015. Such arrests carry a record for individuals that may be highly detrimental to their future. The high level of arrests despite major changes in prohibition laws suggests that there may be a reluctance to forsake traditional behavior and that there is possibly a strong vested interest in the process of criminalizing marijuana—an industry in its own right.

The situation in Europe is mixed, with countries such as the Netherlands extending their long-held policy of partial legalization of cannabis by introducing national well-regulated sources of supply. Other countries, such as Britain, cling to the policy of treating cannabis as an illegal Schedule I drug, although cannabis use is common and the police are becoming more lenient in their treatment of cannabis possession. Nevertheless, the substantial illegal market is completely without any quality control and is served largely by growers with connections to criminal organizations, using indoor cultivation techniques often involving "slave" labor in the form of Vietnamese young people who are illegal immigrants. There is little sign of change in the United Kingdom, although the Liberal Democrat Party did include a relaxation of marijuana laws in its manifesto for the 2016 election, and there are active campaigns promoting a relaxation of the prohibition laws (Beckley Foundation, 2012; All-Party Parliamentary Group for Drug Policy Reform, 2016; Liberal Democrat Party UK, 2016).

Conclusion

Cannabis research is flourishing, despite the difficulties that scientists have in accessing high-quality cannabis. However, many questions remain: Can new medicines be discovered and developed based on the current knowledge of the biosynthesis, actions, and inactivation of endocannabinoids? Can genetic screening identify people who are particularly susceptible to cannabis use disorder and possibly to psychosis? Can researchers pinpoint in more detail how endocannabinoids modulate neural activity and how they change on exposure to stress?

Scientific research will tackle all these questions and more in the coming decades. Meanwhile, we should remember the caution voiced by Volkow et al. (2015):

> The changing landscape of cannabis use (e.g., strains with higher THC potency, new routes of administration ["vaping" and edibles], and novel drug combinations, and a culture of rapidly changing norms and perceptions raise the possibility that our current, limited knowledge may only apply to the ways in which the drug was used in the past.

REFERENCES

Abel EL. *Marihuana: The First Twelve Thousand Years*. New York: Plenum, 1943. (Reprinted 1980)

Abood ME, Martin BR. Neurobiology of marijuana abuse. *Trends Pharmacol Sci*. 1992;13:201–206.

Abrahamov A, Abrahamov A, Mechoulam R. An efficient new cannabinoid antiemetic in pediatric oncology. *Life Sci*. 1995;56:2097–2102.

Adams R. Marihuana. *Harvey Lect*. 1941–1942;37:168–197.

Adams R. Marihuana. *Harvey Lect*. 1942;18:705–730.

Adams IB, Martin BR. Cannabis: Pharmacology and toxicology in animals and humans. *Addiction*. 1996;91:1585–1614.

Advisory Committee on Drug Dependence. *Cannabis*. London: Her Majesty's Stationery Office; 1969.

Advisory Council on the Misuse of Drugs. 2009, "Consideration of the major cannabinoid agonists," July 16; 2012, "Further consideration of the synthetic cannabinoids," October 18; 2014, "Advice on 'third generation' synthetic cannabinoids," March 11; 2016, "Third generation synthetic cannabinoids, updated," December 12 (https://www.government UK /organisations/advisory-cuncil-on-the-misuse-of-dugs).

Agarwal N, Pacher P, Tegeder I, et al. Cannabinoids mediate analgesia largely via peripheral type 1 cannabinoid receptors in nociceptors. *Nature Neurosci*. 2007;10:870–879.

Agurell S, Halldin M, Lindgren J-E, et al. Pharmacokinetics and metabolism of Δ^1-tetrahydrocannabinol and other cannabinoids with emphasis on man. *Pharmacol Rev*. 1986;38:21–38.

Alonso M, Serrano A, Vida M, et al. Anti-obesity efficacy of LH-21, a cannabinoid CB1 receptor antagonist with poor brain penetration, in diet-induced obese rats. *Br J Pharmacol*. 2012;165:2274–2291.

Ambra TE, Eissenstat MA, Ab J, et al. C-attached aminoalkylindoles: Potent cannabinoid mimetics. *Bioorg Med Chem Lett.* 1996;6:17–22.

American Medical Association. Report of the Council on Scientific Affairs to AMA House of Delegates on Medical Marijuana. CSA Report I-97; 1997.

American Psychiatric Association. *Diagnostic and Statistical Manual of Mental Disorders.* 5th ed. Washington, DC: American Psychiatric Association; 2013.

Andreae MH, Carter GM, Shaparin N, et al. Inhaled cannabis for chronic neuropathic pain: A meta-analysis of individual patient data. *J Pain.* 2015;16:1221–1232.

Andreasson S, Allebeck P, Engstrom A, Rydberg U. Cannabis and schizophrenia: A longitudinal study of Swedish conscripts. *Lancet.* 1987;2:1483–1485.

Andreasson S, Allebeck P, Rydberg U. Schizophrenia in users and nonusers of cannabis. *Acta Psychiatr Scand.* 1989;79:505–510.

Anslinger H, Cooper CR. Marihuana: Assassin of youth. *American Magazine,* July 1937, p. 150.

Asbridge M, Hayden JA, Cartwright JL. Acute cannabis consumption and motor vehicle collision risk: Systematic review of observational studies. *Br Med J.* 2012;344:e536.

Association of Public Health Laboratories. Guidance for state medical cannabis testing programs. Silver Spring, MD: Association of Public Health Laboratories; May 2016.

Auer R, Vittinhoff E, Yaffe K, et al. Association between lifetime marijuana use and cognitive function in middle age. *JAMA Intern Med.* 2016;176:352–361.

Axelrod J, Felder C. Cannabinoid receptors and their endogenous agonist anandamide. *Neurochem Res.* 1998;23:575–581.

Aydelotte JD, Brown LH, Luftman KM, et al. Crash fatality rates after recreational marijuana legalization in Washington and Colorado. *Am J Public Health.* 2017;107:1329–1331.

Bab I, Alexander S. (eds.). Special issue: Cannabinoids in biology and medicine. *Br J Pharmacol.* 2011;163:1327–1562.

Babson KA, Sottile J, Morabito B. Cannabis, cannabinoids and sleep. *Curr Psychiatry Rep.* 2017;19(4):23.

Barnes MP. Sativex: Clinical efficacy and tolerability in the symptoms of multiple sclerosis and neuropathic pain. *Expert Opin Pharmacother.* 2006;7:607–615.

Barnes MP. Cannabis—The evidence for medical use. All Party Parliamentary Group enquiry into medical cannabis; 2016. https://drive.googlecom/file/d/0B0c_8hkDJu0DUDZMUzhoY1RqMG8/view.

Beal JE, Olson R, Laubenstein L, et al. Dronabinol as a treatment for anorexia associated with weight loss in patients with AIDS. *J Pain Symptom Management.* 1995;10:89–97.

Beckley Foundation. Cannabis policy: Moving beyond stalemate. Oxford, UK: Beckley Foundation; 2012.

Belendiuk KA, Baldini LL, Bonn-Miller MO. Narrative review of the safety and efficacy of marijuana for the treatment of commonly state-approved medical and psychiatric disorders. *Addict Sci Clin Pract.* 2015;10:10.

Bell R, Wechsler H, Johnston LD. Correlates of college student marijuana use: Results of a US national survey. *Addiction.* 1997;92:571–581.

Bellochio L, Mancini G, Vincenatti V, et al. Cannibinoid receptors as therapeutic targets for obesity and metabolic diseases. *Curr Opin Pharmacol.* 2006;6:586–591.

Bergland C. Cannabis use disorders are escalating *Psychology Today,* March 21, 2016.

Berke J, Hernton C. *The Cannabis Experience.* Aylesbury, UK: Hazell Watson & Viney; 1974. (Reprinted by Quartet Books, London; 1977)

Bifulco M, Pisanti S. Medicinal use of cannabis in Europe. *EMBO Rep.* 2015;16(2):e201439742.

Bonnet U, Preuss UW. The cannabis withdrawal syndrome: Current insights. *Subst Abuse Rehabil.* 2017;8:9–37.

Booth M. *Cannabis: A History.* London: Transworld Publishers; 2003.

Bostan AC, Strick PL. The cerebellum and basal ganglia are interconnected. *Neuropsychol Rev.* 2010;20:261–270.

Braude MC. Toxicology of cannabinoids. In: Paton WM and Crown J (eds.), *Cannabis and Its Derivatives.* Oxford, UK: Oxford University Press; 1972: 89–99.

British Crime Survey. Drug misuse: Findings from the 2015 to 2016 CSEW second edition. 2016. https://www.gov.uk/government/statistics/drug-misuse-findings-from-the-2015-to-2016-csew.

British Medical Association. *Therapeutic Uses of Cannabis.* Reading, UK: Harwood Academic; 1997.

Broyd SJ, van Hell HH, Beale C, Yücel M, Solowij N. Acute and chronic effects of cannabinoids on human cognition—A systematic review. *Biol. Psychiatry.* 2016;79:557–567.

Bruijnzeel AW, Qi X, Guzhva LV, et al. Behavioral characterization of the effects of cannabis and anandamide in rats. *PLoS One.* 2016;11(4):e0153327.

Budney AJ, Hughes JR, Moore BA, Vandrey V. Review of the validity and significance of cannabis withdrawal syndrome. *Am J. Psychiatry.* 2004;161:1967–1977.

Burston JJ, Sagar DR, Shao P, et al. Cannabinoid CB_2 receptors regulate central sensitization and pain responses associated with osteoarthritis of the knee joint. *PLoS One.* November 25, 2013.

Butterfield D. Multiple sclerosis: Here's why cannabis is so effective against MS. *Herb.* July 28, 2016. http://herb.co/2016/07/28/marijuana-and-ms.

Butterfield D. How cannabis successfully combats PTSD. *Herb.* April 23, 2017. http://herb.co/2017/04/23/cannabis-combats-ptsd.

Campbell FA, Tramer M, Carroll D, et al. Are cannabinoids an effective and safe treatment option in the management of pain? A qualitative systematic review. *BMJ.* 2001;323:13.

Canadian Government. Legalizing and strictly regulating cannabis: The facts. April 13, 2017 https://www.canada.ca/en/services/health/campaigns/legalizing-strictly-regulating-cannabis-facts.html.

Carai M, Colombo G, Maccioni P, Gessa GL. Efficacy of rimonabant and other CB_1 receptor antagonists in reducing food intake and body weight: Preclinical and clinical data. *CNS Drug Rev.* 2006;12:91–99.

Carlson G, Wang Y, Alger BE. Endocannabinoids facilitate the induction of LTP in the hippocampus. *Nature Neurosci.* 2002;5:723–724.

Cascio MG, Pertwee RG. Known pharmacological actions of nine nonpsychotropic phytocannabinoids. In: Pertwee RG (ed.), *Handbook of Cannabis.* Oxford, UK: Oxford University Press; 2015: 137.

Castillo PE, Younts TJ, Chavez A, Hashimotodan Y. Endocannabinoid signaling and synaptic function. *Neuron* 2012;76:70–81.

Castle DJ, Murray R (eds.). *Marijuana and Madness.* Cambridge, UK: Cambridge University Press; 2004.

Caulkins JB, Kilmer B, Kleiman AR. *Marijuana Legalization: What Everyone Needs to Know.* 2nd ed. Oxford, UK: Oxford University Press; 2016.

Chait LD. Delta-9-tetrahydrocannabinol content and human marijuana self-administration. *Psychopharmacology.* 1989;98:51–55.

Chalsma AL, Boyum D. *Marijuana Situation Assessment.* Washington, DC: Office of National Drug Control Policy; 1994: 5.

Chan PC, Sills RC, Braun AG, Haseman JK, Bucher JR. Toxicity and carcinogenicity of delta-9-tetrahydrocannabinol in Fischer rats and B6C3F1 mice. *Fundam. Appl Toxicol.* 1996;30:109–117.

Chopra IC, Chopra RN. The use of cannabis drugs in India. *Bull. Narc.* 1957;January:4–29.

Clarke RC. *Marijuana Botany.* Berkeley, CA: Ronin; 1981.

Clifford DB. Tetrahydrocannabinol for tremor in multiple sclerosis. *Ann Neurol.* 1983;13:669–671.

Cluny NL, Vemuri VK, Chambers AP, et al. A novel peripherally restricted cannabinoid receptor antagonist, AM6545, reduces food intake and body weight, but does not cause malaise, in rodents. *Br J Pharmacol.* 2010;161:629–642.

Compton DR, Rice KC, De Costa BR, et al. Cannabinoid structure–activity relationships: Correlation of receptor binding and in vivo activities. *J Pharmacol Exp Ther.* 1993;265:218–226.

Compton RP, Berning A. Drug and alcohol crash risk (Traffic Safety Facts Research Note, Report No. DOT HS 812 117). Washington, DC: National Highway Traffic Safety Administration; 2015.

Compton WM, Han B, Jones CM, et al. Marijuana use and use disorders in adults in the USA, 2002–14: Analysis of annual cross-sectional survey. *Lancet Psychiatry.* 2016;3:954–964.

Consroe P, Musty R, Rein J, Tillery W, Pertwee R. The perceived effects of smoked cannabis on patients with multiple sclerosis. *Eur Neurol.* 1996;38:44–48.

Costa B, Comelli F. Pain. In: Pertwee R (ed.), *Handbook of Cannabis.* Oxford, UK: Oxford University Press; 2015: 473–501.

Craig T. In the land of towering pot plants farmers brace for a buzz-kill. *Washington Post,* September 24, 2015. https://www.washingtonpost. com/world/in-the-land-of-towering-pot-plants-pakistani-farmers-brace-for-a-buzz-kill/2015/09/22/01d9d708-5d79-11e5-8475-781cc9851652_story.html?utm_term=.7ca75cdd4592.

Cravatt BF, Demarest K, Patricelli MP, et al. Supersensitivity to anandamide and enhanced endogenous cannabinoid signaling in mice lacking fatty acid amide hydrolase. *Proc Natl Acad Sci USA.* 2001;98:9371–9376.

Crean RD, Crane NA, Mason BJ. An evidence based review of acute and long term effects of cannabis use on executive cognitive function. *J Addict Med.* 2011;5:1–8.

Crippa JH, Zuardi AW, Martin-Santos R. Cannabis and anxiety: A critical review. *Hum Psychopharmacol Clin Exp.* 2009;24:515–523.

De Fonseca FR, Carrera MRA, Navarro M, Koob GF, Weiss F. Activation of corticotropin-releasing factor in the limbic system during cannabinoid withdrawal. *Science.* 1997;276:2050–2054.

De Luca MA, Bimpisidis Z, Melis M, et al. Stimulation of in vivo dopamine transmission and intravenous self-administration in rats and mice by JWH-018, a Spice cannabinoid. *Neuropharmacology.* 2015;99:705–714.

De Luca MA, Castelli MP, Loi B, et al. Native CB1 receptor affinity, intrinsic activity and accumbens shell dopamine stimulant properties of third generation SPICE/K2 cannabinoids: BB-22, SF-BB-22, 5F-AKB-48 and STS-135. *Neuropharmacology.* 2016;105:630–638.

Deng L, Guidon J, Cornett B, et al. Chronic cannabinoid receptor 2 activation reverses paclitaxel neuropathy without tolerance or cannabinoid receptor 1-dependent withdrawal. *Biol Psychiatry.* 2015;77:475–487.

Denton TF, Earleywine M. Pothead or pot smoke? A taxometric investigation of cannabis dependence. *Subst Abuse Treat Prev Policy.* 2006;1:22.

Devane WA, Dysarz A, Johnson MR, Melvin LS, Howlett A. Determination and characterization of a cannabinoid receptor in rat brain. *Mol Pharmacol.* 1988;34:605–613.

Devinsky O, Cilio MR., Cross H, et al. Cannabidiol: Pharmacology and potential therapeutic role in epilepsy and other neuropsychiatric disorders. *Epilepsia.* 2014;55:791–802.

Devinsky O, Cross JH, Laux L, et al. Trial of cannabidiol for drug-resistant seizures in the Dravet syndrome. *N Engl J Med.* 2017;377:699–700.

Dhopeshwarkar A, Mackie K. CB2 cannabinoid receptors as a therapeutic target: What does the future hold? *Mol Pharmacol.* 2014;85:430–437.

Di S, Itoga CA, Fisher MO, et al. Acute stress suppresses synaptic inhibition and increases anxiety via endocannabinoid release in the basolateral amygdala. *J Neurosci.* 2016;36:8461–8470.

Di Marzo V, De Petrocellis L, Bisogno T. The biosynthesis, fate and pharmacological properties of endocannabinoids. *Handb Exp Pharmacol.* 2005;168:147–185.

Di Marzo V, Melck D, Bisogno T, De Petrocellis L. Endocannabinoids: Endogenous cannabinoid receptor ligands with neuromodulatory function. *Trends Neurosci.* 1998;21:521–528.

Dixon WE. The pharmacology of cannabis indica. *Br Med J.* 1899, November 11:1354–1357.

Doll R, Peto R, Boreham J, Sutherland I. Mortality from cancer in relation to smoking: 50 years' observations on British doctors. *Br J Cancer.* 2005;92:426–429.

Drugs.com. Medications for nausea/vomiting. 2016. https://www.drugs.com/condition/nausea-vomiting.html.

D'Souza DC, Radhakrishnan R. Medical marijuana. Is the cart before the horse? *JAMA.* 2015;313:2431–2432.

D'Souza DC, Radhakrishnan R, Sherif M, et al. Cannabinoids and psychosis. *Curr Pharm Des.* 2016;22:6380–6391.

Earlywine M. *Understanding Marijuana.* Oxford, UK: Oxford University Press; 2002.

Egerton A, Allison C, Brett RR, Pratt JA. Cannabinoids and prefrontal cortical function: Insights from preclinical studies. *Neurosci Biobehav Rev.* 2006;30:680–695.

Egertova M, Elphick MR. Localization of cannabinoid receptors in the rat brain using antibodies to the intracellular C-terminal tail of CB1. *J Comp Neurol.* 2000;422:159–171.

Elphick MR, Egertova M. The neurobiology and evolution of cannabinoid signalling. *Philos Trans R Soc London.* 2001;356:381–408.

ElSohly M, Gul W. Constituents of *Cannabis sativa.* In: Pertwee RG (ed.), *Handbook of Cannabis.* Oxford, UK: Oxford University Press; 2015: 1–22.

ElSohly MA, Gul W, Wanas AS, Radwan MM. Synthetic cannabinoids: Analysis and metabolites. *Life Sci.* 2014;97(1):78–90.

ElSohly M, Mahomood R (eds.). *Marijuana and the Cannabinoids.* New York: Springer; 2010.

Emrich HM, Leweke FM, Schneider U. Towards a cannabinoid hypothesis of schizophrenia: Cognitive impairments due to dysregulation of the endogenous cannabinoid system. *Pharmacol Biochem Behav.* 1997;56: 8030–8080.

Englund A, Stone M, Morrison PD. Cannabis in the arm: What can we learn from intravenous cannabinoid studies? *Curr Pharm Des.* 2012; 18:4906–4914.

Enserink M. More details emerge on fateful French drug trial. *Science.* January 16, 2016.

Erowid. Cannabis testing. n.d. http://www.erowid.org/plants/cannabis/cannabis_testing.shtml.

European Monitoring Centre for Drugs and Drug Addiction. Synthetic cannabinoids in Europe. *Perspective on Drugs.* May 31, 2016a.

European Monitoring Centre for Drugs and Drug Addiction. Statistical bulletin. 2016b. http://www.emcdda.europa.eu/data/stats2016.

European Monitoring Centre for Drugs and Drug Addiction. United Kingdom, country drug report 2017. 2017. http://www.emcdda.europa.eu/publications/country-drug-reports/2017/united-kingdom_en.

Family Council. Number of deaths caused by marijuana much more than 0. 2016. https://familycouncil.org/?p=11795.

Felder CC, Glass M. Cannabinoid receptors and their endogenous agonists. *Annu Rev Pharmacol Toxicol.* 1998;38:179–200.

Fernández-Ruiz J. The endocannabinoid system as a target for the treatment of motor dysfunction. *Br J Pharmacol.* 2009;156:1029–1040.

Fernández-Ruiz J, Sagredo O, Pazos MR, et al. Cannabidiol for neurodegenerative disorders: Important new clinical applications for this phytocannabinoid. *Br J Clin Pharmacol.* 2013;75:323–333.

Fields HL, Meng ID. Watching the pot boil: Selective antagonists of the two cannabinoid receptors unveil distinct but synergistic peripheral analgesic activities for endogenous cannabinoids. *Nature Med.* 1998;4:1000–1009.

Finnerup NB. Treatment of neuropathic pain: Opioids, cannabinoids, and topical agents. Paper presented at the 15th World Congress on Pain, 2014.

Fong TM, Heymsfield SB. Cannabinoid-1 receptor inverse agonists: Current understanding of mechanism of action and unanswered questions. *Int J Obesity.* 2009;33:947–955.

Fontes MA, Bolla KI, Cunha PJ, et al. Cannabis use before age 15 and subsequent executive functioning. *Br J Psychiatry.* 2011;198:442–447.

Fowler C. Transport of endocannabinoids across the plasma membrane and within the cell. *FEBS J.* 2013;280:1895–1904.

Freedman D, Dunn G, Evans N, et al. How cannabis causes paranoia: Using the intravenous administration of Δ^9-tetrahydrocannabinol (THC) to identify key cognitive mechanisms leading to paranoia. *Schizophren Bull.* 2015;41:391–399.

French ED, Dillon K, Wu X. Cannabinoids excite dopamine neurons in the ventral tegmentum and substantia nigra. *NeuroReport.* 1997;8:649–652.

Fride E, Gobshtis N, Dahan H, et al. The endocannabinoid system during development: Emphasis on perinatal events and delayed effects. *Vitam Horm.* 2009;81:139–158.

Fried PA. Prenatal exposure to tobacco and marijuana: Effects during pregnancy, infancy, and early childhood. *Clin Obstet Gynecol.* 1993;36:319–337.

Fried PA, Watkinson B, Gray R. Differential effects on cognitive functioning in 13–16-year-olds prenatally exposed to cigarettes and marijuana. *Neurotoxicol Teratol.* 2003;25:427–436.

Gage SH, Hickman M, Zamit S. Association between cannabis and psychosis: Epidemiologic evidence (review). *Biol Psychiatry.* 2016;79:549–556.

Gallup. One in eight U.S. adults say they smoke marijuana. 2016. http://www.gallup.com/poll/194195/adults-say-smoke-marijuana.aspx.

Gaoni Y, Mechoulam R. Isolation, structure, and partial synthesis of an active constituent of hashish. *J Am Chem Soc.* 1964;86:1646–1647.

Geigling I, Hosak L, Mossner R, et al. Genetics of schizophrenia: A consensus paper of the WFSBP Task Force on Genetics. *World J Biol Psychiatry.* 2017;18:492–505.

Gettman J. Marijuana production in the United States (2009). *Bulletin of Cannabis Reform,* November. http://www.drugscience.org/bcr/index. html.

Ghodse H. When too much caution can be harmful. *Addiction.* 1996; 91:764–766.

Gilchrist K. Netherlands votes in favour of regulated marijuana production. CNBC. February 22, 2017. https://www.cnbc.com/2017/02/22/ netherlands-votes-in-favour-of-regulated-marijuana-production.html.

Giuffrida A, Parsons LH, Kerr TM, de Fonseca FR, Navarro M, Piomelli D. Dopamine activation of endogenous cannabinoid signalling in dorsal striatum. *Nature Neurosci.* 1999;2:358–363.

Gizer R, Bizon C, Gilder DA, Ehlers CL, Wilhelmsen KC. Whole genome sequence study of cannabis dependence in two independent cohorts. *Addict Biol.* 2018;23:461–473.

Gong J-P, Onaivi ES, Ishiguro H, et al. Cannabinoid CB2 receptors: Immunohistochemical localization in rat brain. *Brain Res.* 2006; 1071:10–23.

Goni U. Uruguay, the first country where you can smoke marijuana wherever you like. *The Guardian,* May 27, 2017.

Gonzalez E. Where does Latin America stand on marijuana legalization? Council of the Americas, April 13, 2017. http://www.as-coa.org/articles/ weekly-chart-where-does-latin-america-stand-marijuana-legalization.

Gonzalez S, Cebeira M, Fernandez-Ruiz J. Cannabinoid tolerance and dependence: A review of studies in laboratory animals. *Pharmacol Biochem Behav.* 2005;81:30–318.

Goode E. *The Marijuana Smokers.* New York: Basic Books; 1970.

Gorelick DA, Levin KH, Copersino ML, et al. Diagnostic criteria for cannabis withdrawal syndrome. *Drug Alcohol Depend.* 2012;123:141–147.

Gorter RW, Butarac M, Cobian EP, van ser Sluis, W. Medical use of camnabis in the Netherlands. *Neurology.* 2005;64:917–919.

Government of Canada. Medical use of cannabis: Information on the new Access to Cannabis for Medical Purposes Regulations. 2016. https:// www.canada.ca/en/health-canada/services/drugs-health-products/ medical-use-marijuana/medical-use-marijuana.html.

Green B, Kavanagh D, Young R. Being stoned: A review of self-reported cannabis effects. *Drug Alcohol Rev.* 2003;22:453–460.

Grether N. In the Netherlands, 38 years of lessons on "tolerating" pot. *America Tonight.* February 21, 2017.

Grinspoon L, Bakalar JB. *Marihuana, the Forbidden Medicine.* New Haven, CT; Yale University Press; 1993. (Revised edition 1997)

Grinspoon L, Bakalar JB. Marijuana as medicine: A plea for reconsideration. *JAMA.* 1995;273:1875–1876.

Gruber SA, Sagar KA, Dahlgren MK, Racine M, Lukas SE. Age of onset of marijuana use and executive function. *Psychol Addict Behav.* 2012;26:496–506.

Grufferman S, Wang HH, DeLong ER, Kimm SYS, Delzell ES, Falletta JM. Environmental factors in the etiology of rhabdomyosarcoma in childhood. *J. Natl. Cancer Inst.* 1982;68:107–113.

Grufferman S, Schwartz AG, Ruymann FB, et al. Parents' use of cocaine and marijuana and increased risk of rhabdomyosarcoma in their children. *Cancer Causes Control.* 1993;4:217–224.

Hall W, Solowij N. Long-term cannabis use and mental health. *Br J Psychiatry.* 1997;171:107–108.

Hamilton I. Here's the truth about whether cannabis causes psychosis—and what you can do to minimise your risk of harm. *The Independent.* April 20, 2017.

Hampson AJ, Grimaldi M, Axelrod J. Cannabidiol and delta-9-tetrahydrocannabinol are neuroprotective antioxidants. *Proc Natl Acad Sci USA* 1998;95:8268–8273.

Han CJ, Robinson JK. Cannabinoid modulation of time estimation in the rat. *Behav Neurosci.* 2001;115:243–246.

Hanus L, Abu-Lafi S, Fride E, et al. 2-Arachidonyl glyceryl ether, and endogenous agonist of the cannabinoid CB1 receptor. *Proc Nat Acad Sci.* 2001;98:3662–3665.

Hasenoehrl C, Storr M, Schicho R. Cannabinoids for treating inflammatory bowel disease: Where are we and where do we go? *Expert Rev Gastroenterol Hepatol.* 2017;11:329–337.

Hasin DS, Kerridge BT, Saha TD, et al. Prevalence and correlates of DSM-5 cannabis use disorder, 2012–2013: Findings from the National Epidemiologic Survey on Alcohol and Related Conditions–III. *Am J Psychiatry.* 2016;173:588–599.

Hasin DS, O'Brien CP, Auriacombe M, et al. DSM-5 criteria for substance use disorders: Recommendations rationale. *Am J Psychiatry.* 2013;170:834–851.

Hayabatbaksh MR, Flemig LS, Gibbons KS, et al. *Pediatric Research.* 2012;71:215–216.

Hazecamp A. An evaluation of the quality of medicinal grade cannabis in the Netherlands. *Cannabinoids.* 2006;1:1–9.

Helping Hands Herbals. 2017. https://helpinghandsdispensary.com.

Herer J. *The Emperor Wears No Clothes*. Newcastle upon Tyne, UK: Green Planet; 1993.

Herkenham M, Lynn AB, Johnson MR, Melvin LS, de Costa BR, Rice KC. Characterization and localization of cannabinoid receptors in rat brain: A quantitative in vitro autoradiographic study. *J Neurosci.* 1991;11:563–583.

Herning RI, Hooker WD, Jones RT. Tetrahydrocannabinol content and differences in marijuana smoking behavior. *Psychopharmacology.* 1986;90:160–162.

Hess C, Schjoeder CT, Pillaiyart T, Madea B, Müller CE. Pharmacological evaluation of synthetic cannabinoids, identified as constituents of spice. *Forensic Toxicol.* 2016;34:329–343.

Heyser S, Hampson RE, Deadwyler SA. Effects of delta-9-tetrahydrocannabinol on delayed match to performance in rats: Alterations in short-term memory associated with changes in task-specific firing of hippocampal cells. *J Pharmacol Exp Ther.* 2016;264:294–307.

Hill KP. Medical marijuana for treatment of chronic pain and other medical and psychiatric problems: A clinical review. *JAMA.* 2015;313:2474–2483.

Hillig KW, Mahlberg PG. A chemotaxonomic analysis of cannabinoid variation in *Cannabis* (Cannabaceae). *Am J Bot.* 2004;91:966–975.

Himmelstein JL. *The Strange Career of Marihuana.* Contributions in Political Science, no. 94. Westport, CT: Greenwood Press; 1978.

Hollister LE. Health aspects of cannabis. *Pharmacol Rev.* 1986;38:1–20.

Hollister LE. Marijuana and immunity. *J Psychoactive Drugs.* 1992;24: 150–164.

Hollister LE. Health aspects of cannabis: Revisited. *Int J Neuropsychopharmacol.* 1998;1:71–80.

Houck JM, Bryan AD, Feldstein Ewing SW. Functional connectivity and cannabis use in high-risk adolescents. *Am J Drug Alcohol Abuse.* 2013;239:414–423.

House of Lords, Select Committee on Science and Technology. *Cannabis—The Scientific and Medical Evidence.* London: The Stationary Office; 1998.

Howlett AC. Cannabinoid receptors (Pert, ignalling). *Handb Exp Pharmacol.* 2005;168:53–80.

Howlett AC, Breivogel CS, Childers SR, et al. Cannabinoid physiology and pharmacology: 30 years of progress. *Neuropharmacology.* 2004;47(Suppl. 1):345–358.

Hua T, Vemura K, Pu M, et al. Crystal structure of the human cannabinoid receptor CB1. *Cell.* 2016;167:750–762.

Huang J, Zhang ZF, Tashkin DP, et al. An epidemiologic review of marijuana and cancer: An update. *Cancer Epidemiol Biomarkers Prev.* 2015;24:15–31.

Huestis MA. Human cannabinoid pharmacokinetics. *Chem Biodivers.* 2007;4:1770–1804.

Huestis MA, Gorelick DA, Heishman DA, et al. Blockade of the effects of smoked marijuana by the CB1-selective cannabinoid receptor antagonist SR141716. *Arch Gen Psychiatry.* 2001;58:322–328.

Huestis MA, Sampson AH, Holicky BJ, Henningfield JE, Cone EJ. Characterization of the absorption phase of marijuana smoking. *Clin Pharmacol Ther.* 1992;52:31–41.

Huddleston T Jr. 5 Companies with the biggest buzz in the marijuana industry. *Fortune.* April 20, 2015. [http://fortune.com/2015/04/20/marijuana-industry-five-companies.

Hughes CE, Stevens A. What can we learn from the Portuguese decriminalization of illicit drugs? *Br J Criminol.* 2010;50:999–1022.

Indian Hemp Drugs Commission. Report. Simla, India: Government Central Printing Office; 1894.

Institute of Medicine; Joy JE, Watson J Jr, Benson JA Jr. (eds.). *Marijuana and Medicine.* Washington, DC: National Academy Press; 1999.

Iseger TA, Bossong MG. A systematic review of the antipsychotic properties of cannabidiol in humans. *Schizophr Res.* 2015;162:153–161.

Iversen LL. *The Science of Marijuana.* New York: Oxford University Press; 2003.

Izzo AA, Borrelli F, Capasso R, Di Marzo V, Mechoulam R. Non-psychotropic plant cannabinoids: New therapeutic opportunities from ancient herb. *Trends Pharmacol Sci.* 2009;30:515–527.

Jahr GHG. *New Homeopathic Pharmacopoeia and Posology or the Preparation of Homeopathic Medicines.* Philadelphia, PA: Dobson; 1842: 137.

Jarbe TUC, McMillan DE. Delta-9-THC as a discriminative stimulus in rats and pigeoms. *Psychopharmacology,* 1980;71:281–289.

Jeffrey DR. Recent advances in treating multiple sclerosis: Efficacy, risks and place in therapy. *Ther Adv Chronic Dis.* 2013;4:45–51.

Johnson FR, Burnell-Nugent M, Lossignol D, et al. Multicenter, double blind, randomized, placebo controlled, parallel-group study of the efficacy, safety, and tolerability of THC–CBD extract and THC extract in patients with intractable cancer-related pain. *J Pain Symptom Manage.* 2010;39:167–179.

Johnston I. As cannabis is widely legalised, China cashes in on an unprecedented boom. *Cannabis Para la Educacion.* January 5, 2014. https://cannabisparalaeducacion.org/patents/china-is-cornering-the-cannabis-patent-market.

Justinova Z, Goldberg SR, Heishman S, Tanda G. Self-administration of cannabinoids by experimental animals and human marijuana smokers. *Pharmacol Biochem Behav.* 2005;81:285–299.

Justinova Z, Tanda G, Redhi GH, et al. Self-administration of delta-9-tetrahydrocannabinol (THC) by drug-naïve squirrel monkeys. *Psychopharmacology.* 2003;169:135–140.

Kair C, Hart CL. Cannabis and psychosis: A critical overview of the relationship. *Curr Psychiatry Rep.* 2016;18:12.

Kandel DB, Davies M. High school students who use crack and other drugs. *Arch Gen Psychiatry.* 1996;53:71–80.

Kano M, Ohno-Shosaku T, Hashimotodani Y, Uchigashima M, Watanabe M. Endocannabinoid-mediated control of synaptic transmission. *Physiol Rev.* 2009;89:309–380.

Kaptchuk TJ, Miller FG. Placebo effects in medicine. *N Engl J Med.* 2015;373:8–9.

Katona I, Sperlagh B, Sik A, et al. Presynaptically located CB1 receptors regulate GABA axon terminals of specific hippocampal interneurons. *J Neurosci.* 1999;19:4544–4558.

Keating GM. Delta-9-tetrahydrocannabinol/cannabidiol oromucosal spray (Sativex): A review in multiple sclerosis-related spasticity. *Drugs.* 2017;77:563–574.

Kemoker JA, Honig EG, Martin GS. The effects of marijuana exposure on expiratory airflow: A study of adults who participated in the US National Health and Nutrition Examination Study. *Ann Am Thorac Soc.* 2015;12:135–141.

Kendall DA, Yudowski G. Cannabinoid receptors in the central nervous system: Their signalling and roles in disease. *Front Cell Neurosci.* 2017;10:294.

Khairy H, Houssen WE. Inactivation of anandamide signaling: A continuing debate. *Pharmaceuticals.* 2010;3:3355–3370.

Kindland. Lawmakers in 4 States Refuse to Give Voters the Pot They Voted For. January 19, 2017.

Knodt M. Contaminated cannabis—The unavoidable consequence of prohibition. *Marijuana News,* January 27, 2017.

Kolodny RC, Masters WH, Kolodner RM, Toro G. Depression of plasma testosterone levels after chronic intensive marihuana use. *N Engl J Med.* 1974;290:872–874.

Korver N, Quee PJ, Boos HB, et al; Genetic Risk and Outcome in Psychosis (GROUP) Investigators. Evidence that familial liability for psychosis is expressed as differential sensitivity to cannabis. *Arch Gen Psychiatry.* 2011;68:138–142.

Kozlowski L, Coambs RB, Ferrence R, et al. Preventing smoking and other drug use. Let the buyers beware and.the interventions be apt. *Can J Public Health.* 1989;80:452–456.

Kozlowsi LT, Wilkinson DA, Skinner W, et al. Comparing tobacco cigarette dependence with other drug dependencies. *JAMA.* 1989;261:898–201.

Kunos G, Tam J. The case for peripheral CB_1 receptor blockade in the treatment of visceral obesity and its cardiometabolic complications. *Br J Pharmacol.* 2011;163:1423–1431.

Kuster JE, Stevenson JI, Ward SJ, D'Ambra TE, Haycock DA. Aminoalkylindole binding in rat cerebellum: Selective displacement by natural and synthetic cannabinoids. *J Pharmacol Exp Ther.* 264:1993;1352–1363.

Ladin DA, Soliman E, Van Dross R. Preclinical and clinical assessment of cannabinoids as anti-cancer agents. *Front Pharmacol.* 2016;7:361.

Lambert DM, Fowler CJ. The endocannabinoid system: Drug targets, lead compounds, and potential therapeutic applications. *J Med Chem.* 2005;48:5059–5087.

Lambert DM, Muccioli GG. Endocannabinoids and related ethanolamines in the control of appetite and energy metabolism emergence of new molecular players. *Curr Opin Clin Nutr Metab Care.* 2007;10:735–744.

Lane SD, Cherek DR, Pietras CJ, Steinberg JL. Performance of heavy marijuana-smoking adolescents on a laboratory measure of motivation. *Addict Behav.* 2005;30(4):815–828.

Langford RM, Mares J, Novotna A, et al. A double-blind, randomized, placebo-controlled, parallel group study of THC/CBD oromucosal spray in combination with the existing treatment regimes, in the treatment of central neuropathic pain in patients with multiple sclerosis. *J Neurol.* 2013;260:984–997.

Lawn W, Freeman TP, Pope RA. Acute and chronic effects of cannabinoids on effort-related decision-making and reward learning: An evaluation of the cannabis "amotivational" hypotheses. *Psychopharmacology.* 2014;233:3537–3552.

Leafy. Cannabis 101: Terpenes—The flavors of cannabis aromatherapy. 2017. https://www.leafly.com.news/cannabis-101/terpenes-the-flavors-of-cannabis-aromatherapy.

Ledent C, Valverde O, Cossu G, et al. Unresponsiveness to cannabinoids and reduced addictive effects of opiates in CB_1 receptor knockout mice. *Science.* 1999;283:401.

Lee Y, Jo D, Chang HY, Pothoulakis C, Im E. Endocannabinoids in the gastrointestinal tract. *Am J Physiol Gastrointest Liver Physiol.* 2016;311:G655.

Lemberger L. Clinical evaluation of cannabinoids in the treatment of disease. In: Harvey DJ, Paton W, Nahas G (eds.), *Marihuana '84*. Oxford, UK: IRL Press; 1985: 673–680.

Levine JD, Gordon NC, Bornstein JC, Fields HL. Role of pain in placebo analgesia. *Proc Natl Acad Sci USA*. 1979;76:3528–3531.

Lewin L. *Phantastica: Narcotic and Stimulating Drugs Their Use and Abuse*. London: Routledge & Kegan Paul; 1931.

Lewis MM, Yang Y, Masilewski E. Chemical profiling of medical cannabis extracts. *ACS Omega*. 2017;2:6091–6103.

Li MH, Suchland KL, Ingram SL. Compensatory activation of cannabinoid CB2 receptor inhibition of GABA release in the rostral ventromedial medulla (RVM) in inflammatory pain. *J Neurosci*. 2017;37:626–636.

Liberal Democrat Party UK. A framework for a regulated market for cannabis in the UK. 2016. https://www.tdpf.org.uk/sites/default/files/A_framework_for_a_regulated_market_for_cannabis_in_the_UK.pdf.

Ligresti A, De Petrocellis M, DiMarzo V. From phytocannabinoids to cannabinoid receptors and endocannabinoids: Pleiotropic, physiological and pathological roles through complex pharmacology. *Physiol Rev*. 2016;96:1593–1659.

Lind J, Enquist M, Ghirlanda S. Animal memory: A review of delayed matching to sample data. *Behav Processes*. 2015;117:52–58.

Lindsey N. A farm in California is going to produce 6,000 pounds of weed per year. *Green Rush Daily*, April 2, 2016. https://www.greenrushdaily.com/farm-california-going-produce-6000-pounds-weed-per-year.

Livni E. A cannabis-business park covering a million square feet is coming to Massachusetts. *Quartz*, December 28, 2016. https://qz.com/872938/the-biggest-marijuana-grow-facility-in-the-us-isnt-where-you-think-it-would-be.

LoVerne J, Duranti A, Tontini A, et al. Synthesis and characterization of a peripherally restricted cannabinoid antagonist, URB447, that inhibits body weight gain in mice. *Bioorganic Med Chem*. 2009;19:639–643.

Lubman DI, Cheetham A, Yücel M. Cannabis and adolescent brain development. *Pharmacol Ther*. 2015;148:1–16.

Ludlow FH. *The Hasheesh Eater: Being Passages from the Life of a Pythagorean*. New York: Harper Bros; 1857.

Luongo L, Starowicz K, Malone S, Di Marzo V. Allodynia lowering induced by cannabinoids and endocannabinoids. *Pharmacol Res*. 2017;119:272–277.

Maccarrone M, Dainese E, Oddi S. Intracellular trafficking of anandamide: New concepts for signalling. *Trends Biochem Sci*. 2010;35:601–608.

Maccarrone M, Guzmán M, Mackie K, et al. Programming of neural cells by (endo)cannabinoids: From physiological rules to emerging therapies. *Nat Rev Neurosci.* 2014;15:786–801.

Maccarrone M, Wenger T. Effects of cannabinoids on hypothalamic and reproductive function. *Handb Exp Pharmacol.* 2005;168:556–571.

Mackie K. Distribution of cannabinoid receptors in the central and peripheral nervous system. *Handb Exp Pharmacol.* 2005;168:299–325.

Mackie K. Cannabinoid receptors as therapeutic targets. *Annu Rev Pharmacol Toxicol.* 2006;46:101–22.

Makriyannis A, Rapaka R. The molecular basis of cannabinoid activity. *Life Sci.* 1990;47:2173–2184.

Malan TP, Ibrahim MM, Lai J, Todd V, Porreca P. CB2 cannabinoid receptor agonists: Pain relief without psychoactive effects? *Curr Opin Pain.* 2003;3:62–67.

Mallet C, Dubray C, Dualé C. FAAH inhibitors in the limelight, but regrettably. *Int J Clin Pharmacol Ther.* 2016;54:498–501.

Manera C, Arena C, Chicca A. Synthetic cannabinoid receptor agonists and antagonists: Implication in CNS disorders. *Recent Pat CNS Drug Discov.* 2016;10:142–156.

Marichal-Cancino BA, Fajarso-Valdèz A, Ruiz-Contreras M, Prosperor-Garcia O. Advances in the physiology of GPR55 in the central nervous system. *Curr Neuropharmacol.* 2017;15:771–778.

Marijuana News. BRICS nations against marijuana and why this is relevant for the whole cannabis world. March 2016. http://420intel.com/articles/2016/03/31/brics-nations-against-marijuana-and-why-relevant-whole-cannabis-world.

Marijuana Policy Group. The economic impact of marijuana legalization in Colorado. October 2016. http://www.mjpolicygroup.com/pubs/MPG%20Impact%20of%20Marijuana%20on%20Colorado-Final.pdf.

Mariscano G, Wotjak CT, Azad SC, et al. The endogenous cannabinoid system controls extinction of aversive memories. *Nature.* 2002;418:488–489.

Marshall CR. The active principle of Indian hemp: A preliminary communication. *Lancet.* 1897;1: 235–238.

Martin A, Rashidian N. *A New Leaf: The End of Cannabis Prohibition.* New York: The New Press; 2017. http://thenewpress.com/books/new-leaf.

Martin BR. Characterization of the antinociceptive activity of Δ9-tetrahydrocannabinol in mice. In: Harvey DJ, Paton W, Nahas G (eds.), "*Marihuana '84.*Oxford, UK: IRL Press; 1985: 685–692.

Martin CS, Chung T, Langenbucher JW. How should we revise diagnostic criteria for substance use disorders in the DSM-V? *J Abnorm Psychol.* 2008;117:561–575.

Martinez JL, Derrick BE. Long-term potentiation and learning. *Annu Rev Psychol.* 1996;47:173–203.

Matias I, Di Marzo V. Endocannabinoids and the control of energy balance. *Trends Endocrinol Metab.* 2007;18:27–37.

Matsuda LA, Lolait SJ, Brownstein MJ, Young AC, Bonner TI. Structure of a cannabinoid receptor and functional expression of the cloned cDNA. *Nature.* 1990;346:561–564.

Matthew RJ, Wilson WH. Acute changes in cerebral blood flow after smoking marijuana. *Life Sci.* 1993;52:757–767.

Matthew RJ, Wilson WH, Coleman RE, Turkington TG, DeGrado TR. Marijuana intoxication and brain activation in marijuana smokers. *Life Sci.* 1997;60:2075–2089.

Matthew RJ, Wilson WH Turkington TG, Coleman RE. Cerebellar activity and time sense after THC. *Brain Res.* 1998;797:183–189.

Mazier W, Saucisse N, Gatta-Cherifi B, Cota D. The endocannabinoid system: Pivotal orchestrator of obesity and metabolic disease. *Trends Endocrinol Metab.* 2015;26:524–537.

McGlothlin WH, West IJ. The marijuana problem: An overview. *Am J Psychiatry.* 1968;125:124–135.

McPartland JM, Duncan M, Di Marzo V, Pertwee RG. Are cannabidiol and deta-9-tetrahydrocannabivarin negative modulators of the endocannabinoid system? A systematic review. *Br J Pharmacol.* 2015;172:737–753.

Mechoulam R. Marihuana chemistry. *Science.* 1970;168:1159–1163.

Mechoulam R, Fride E, DiMarzo V. Endocannabinoids. *Eur J Pharmacol.* 1998;108:1–13.

Mechoulam R, Hanus L. A historical overview of chemical research on cannabinoids. *Chem Phys Lipids.* 2000;108:1–13.

Mechoulam R, Hanuš LO, Pertwee R, Howlett AC. Early phytocannabinoid chemistry to endocannabinoids and beyond. *Nat Rev Neurosci.* 2014;15:757–764.

Melges FT, Tinklenberg JR, Hollister LE, Gillespie HK. Marihuana and the temporal span of awareness. *Arch. Gen. Psychiatry.* 1971;24:564–567.

Mic. Where is marijuana legal in the United States? List of Recreational and Medical States. 2017 https://mic.com/articles/126303/where-is-marijuana-legal-in-the-united-states-list-of-recreational-and-medicinal-states#.Z73aM0Dill01.

Moon JB. Sir William Brooke O'Shaugnessy—The foundations of fluid therapy and the Indian telegraph service. *N Engl J Med.* 1967;276:283–284.

Moore TH, Zammit S, Lingford-Hughes A, et al. Cannabis use and risk of psychotic or affective mental health outcomes: A systematic review. *Lancet.* 2007;370:319–328.

Morena M, Patel S, Bains JS, Hill MN. Neurobiologocal interactions between stress and the endocannabinoid system. *Neuropsychopharmology.* 2016;41:80–102.

Morgan CJA, Freeman TP, Powell J, Curran HV.2017 AKT1 genotype moderates the acute psychotomimetic effects of naturalistically smoked cannabis in young cannabis smokers. *Transl Psychiatry.*;6:e738.

Morrison PD, Zois V, McKeown DA, et al. The acute effects of intravenous delta-9-tetrahydrocannabinol on psychosis, mood and cognitive functioning. *Psychol. Med.* 2009;39:1607–1616.

Morton GJ, Cummings DE, Baskin DG, et al. Central nervous system control of food intake and body weight. *Nature.* 2006;443:289–295.

Mücke M, Phillips T, Radbruck L. Cannabinoids for chronic neuropathic pain. *Cochrane Rev.* December 20, 2016:CD012182.

Murray RM, Di Forti M. Cannabis and psychosis: What degree of proof do we require? *Biol Psychiatry.* 2016;79:514–515.

Murray RM, Quigley H, Quattrone D, Englund A, Di Forti M. Traditional marijuana, high-potency cannabis and synthetic cannabinoids: Increasing risk for psychosis. *World Psychiatry.* 2016;15:195–204.

NarcoCheck. Marijuana. 2015. http://www.narcocheck.com/en/saliva-drug-tests/thc-marijuana-saliva-test.html.

National Academies of Sciences, Engineering, and Medicine. *The Health Effects of Cannabis and Cannabinoids: The Current State of Evidence and Recommendations for Research.* Washington, DC: National Academies Press; 2017.

National Institute for Clinical Excellence. Multiple sclerosis in adults: Management. Clinical guideline CG186. October 2014.

National Institute on Drug Abuse. Research reports: Drugged driving. Revised June 2016a.

National Institute on Drug Abuse. Marijuana. 2016b. https://www.drugabuse.gov/infofacts/marijuana.html.

National Institute on Drug Abuse. Is marijuana a gateway drug? 2017. https://www.drugabuse.gov/publications/research-reports/marijuana/marijuana-gateway-drug.

National Multiple Sclerosis Society. Medications. 2017. https://www.nationalmssociety.org/Treating-MS/Medications.

National Organization for Reform of the Marijuana Laws. Marijuana stocks to watch in 2017 & beyond. February 23, 2017. http://marijuanastocks. com/marijuana-stocks-to-watch-in-2017-beyond.

National Programme on Substance Abuse Deaths. Annual report. 2017. [npsad@sgul.ac.uk]

Negrete JC, Knapp WP, Douglas DE, Smith WB. Cannabis affects the severity of schizophrenic symptoms: Results of a clinical survey. *Psychol Med.* 1986;16:515–520.

New York Times. Six powerful reasons to legalize marijuana [Editorial]. 2015.

Nicolussi S, Deutsch J. Endocannabinoid transport revisited. *Vitam Horm.* 2015;98:441–485.

Niesink, RJM, van Laar MW. Does cannabidiol protect against adverse psychological effects of THC? *Front Psychiatry.* 2013;4:130–144.

Numikko TJ, Serpell MG, Hoggart B, et al. Sativex successfully treats neuropathic pain characterised by allodynia: A randomised, double-blind, placebo controlled clinical trial. *Pain.* 2007;133:210–220.

O'Connell BK, Gloss D, Devinsky O. Cannabinoids in treatment-resistant epilepsy: A review. *Epilepsy Behav.* 2017;70:341–348.

Office for National Statistics. Deaths related to drug poisoning, England and Wales. 2017. https://www.ons.gov.uk/peoplepopulationandcommunity/birthsdeathsandmarriages/deaths/datasets/deathsrelatedtodrugpoisoningenglandandwalesreferencetable.

Office of Medicinal Cannabis, Netherlands. The Dutch medicinal cannabis program. 2016. http://www.ncsm.nl/english/the-dutch-medicinal-cannabis-program.

Oka S, Ota R, Shima M, Yamashita A, Suguira T. GPR55 is a novel lysophosphatidic acid receptor. *Biochem Biophys Res Commun.* 2010;395:232–237.

Oleson EB, Cheer JF. A brain on cannabinoids: The role of dopamine release in reward seeking. *Cold Spring Harb Perspect Med.* 2012;2:a012229.

Orr C, Morioka R, Behan B, et al. Altered resting-state connectivity in adolescent cannabis users. *Am J Drug Alcohol Abuse.* 2013;39:372–381.

Osei-Hyiamin D, DePetrillo M, Pacher P, et al. Endocannabinoid activation of hepatic CB1 receptors stimulates fatty acid synthesis and contributes to diet-induced obesity. *J Clin Invest.* 2005;115:1298–1305.

O'Shaughnessey WB. On the preparation of the Indian hemp, or gunjah (*Cannabis indica*) and their effects on the animal system in health and their utility in the treatment of tetanus and other convulsive disorders. *Trans. Med Phys Soc Calcutta.* 1842;8:421–461.

O'Sullivan SE. Phytocannabinoids and the cardiovascular system. In: Pertwee R (ed.), *Handbook of Cannabis*. Oxford, UK: Oxford University Press; 2015: 208–226.

Pacher P, Batkai S, Kunos G. Cardiovascular pharmacology of cannabinoids. *Handb Exp Pharmacol*. 2005;168:600–625.

Patel S, Hill MN, Cheer JF, Wotiak CT, Holmes A. The endocannabinoid system as a target for novel anxiolytic drugs. *Neurosci Behav Res*. 2017;76:56–66.

Paule MG, Allen RR, Bailey JR, et al. Chronic marijuana smoke exposure in the rhesus monkey, II: Effects on progressive ratio and conditioned position responding. *J Pharmacol Exp Ther*. 1992;260:210–222.

Pério A, Rimaldi-Carmona M, Maruani J, Barth F, Le fur G, Soubrié P. Central mediation of the cannabinoid cue: Activity of a selective CB1 antagonist, SR 141716A. *Behav. Pharmacol*. 1996;7:65–71.

Pert CB, Snyder SH. Opiate receptor demonstration in nervus tissue. *Science*. 1973;179:1011–1014.

Pertwee R. Tolerance to and dependence on psychotropic cannabinoids. In: Pratt J (ed.), *The Biological Basis of Drug Tolerance*. London: Academic Press; 1991: 231–265.

Pertwee R. Inverse agonism and neutral antagonism at cannabinoid CB1 receptors. *Life Sci*. 2005;76:1307–1324.

Pertwee R. Cannabinoid receptor ligands. Tocris Scientific Reviews. 2006. Accessed October 3, 2017. http://www.tocris.com/scientific-reviews.

Pertwee R. Receptors and channels targeted by synthetic cannabinoid receptor agonists and antagonists. *Curr Med Chem*. 2010;17:1–27.

Pertwee R. Endocannabinoids and their pharmacological actions. *Handb Exp Pharmacol*. 2015b;231:1–37.

Peto R. Influence of dose and duration of smoking on lung cancer rates. In: Zaridze DG, Peto R (eds.), *Tobacco: A Major International Health Hazard*. Publication No. 74. Lyon, France: International Agency for Research on Cancer; 1986: 23–33.

Peto R, Lopez AD, Boreham J, Thun M, Heath C Jr, Doll R. Mortality from smoking worldwide. *Br Med Bull*. 1996;52:12–21.

Piomelli D. The molecular logic of endocannabinoid signalling. *Nat Rev Neurosci*. 2003;4:873–840.

Piomelli D, Tarzia G, Duranti A, et al. Pharmacological profile of the selective FAAH inhibitor KDS-4103 (URB597). *CNS Drug Rev*. 2006;12: 21–28.

Plasse TF, Gorter RW, Krasnow SH, Lane M, Shepard KV, Wadleigh RG. Recent clinical experience with dronabinol. *Pharmacol Biochem Behav*. 1991;40:695–700.

Polen MR, Sidney S, Tekawa IS, Sadler M, Friedman GD. Health care use by frequent marijuana smokers who do not smoke tobacco. *West J Med.* 1993;158:596–601.

Police Foundation. *Drugs and the Law.* London: Police Foundation; 2000.

Portenoy RG, Ganae-Mortan ED, Allende S, et al. Nabiximols for opioid-treated cancer patients with poorly controlled chronic pain: A randomized, placebo-controlled graded-dose trial. *J Pain.* 2012;13:438–449.

Potter D. Cannabis horticulture. In: Pertwee R (ed.), *Handbook of Cannabis.* Oxford, UK: Oxford University Press; 2014: 23–43.

Preuss JW, Watzke AB, Zimmermann J, et al. Cannabis withdrawal severity and short-term course among cannabis dependent adolescents and young adult patients. *Drug Alcohol dependence.* 2010;106:133–141.

ProCon.org 2017. 28 legal medical marijuana states and DC. 2016. https:// medicalmarijuana.procon.org/view.resource.php?resourceID=000881.

Prospéro-Garcia O, Amancia-Belmont O, Becerril Melendez AL, Ruiz-Contreras AE, Diaz M. Endocannabinoids and sleep. *Neurosci Biobehav Rev.* 2016;71:671–679.

Pryce G, Visintin C, Ramagopalan SV, et al. Control of spasticity in a multiple sclerosis model using central-nervous system-excluded CB1 cannabinoid receptor agonists. *FASEB J.* 2014;28:117–130.

Ramesh D, Schlosburg JE, Lichtman AH. Marijuana dependence: Not just smoke and mirrors. *ILAR J.* 2011;52: 295–308.

Reidel G, Davies SN. Cannabinoid function in learning, memory and plasticity. *Handb Ex Pharmacol.* 2005;168:445–477.

Reynolds JR. On the therapeutic uses and toxic effects of *Cannabis indica. Lancet* 1890;1:637–638.

Riebe CJ, Wotjak CT. Endocannabinoids and stress. *Stress.* 2011;14:384–397.

Robbe H. Marijuana's impairing effects on driving are moderate when taken alone, but severe when combined with alcohol. *Human Psychopharmacol.* 1998;13:S70–S78.

Robbins TW, James M, Owen AM, Sahakian BJ, McInnes L, Rabbitt P. Cambridge Neuropsychological Test Automated Battery (CANTAB): A factor analytic study of a large sample of normal elderly volunteers. *Dementia.* 1994;5:266–81.

Roberts C. Thanks to "dabbing," it is possible to overdose marijuana. *SF Weekly News,* March 13, 2013.

Robinson R. *The Great Book of Hemp.* Rochester, VT: Park Street Press; 1996.

Roffman R, Stephens S (eds.). *Cannabis Dependence: Its Nature, Consequences and Treatment.* International Research Monographs in Addiction. Cambridge, UK: Cambridge University Press; 2006.

Rosenberg EC, Patra PH, Whalley BJ. Therapeutic effects of cannabinoids in animal models of seizures, epilepsy, epileptogenesis, and epilepsy-related neurodegeneration. *Epilepsy Behav.* 2017;70:319–327.

Rough L. Leafly's state-by-state guide to cannabis testing regulations. 2016. https://www.leafly.com/news/industry/leaflys-state-by-state-guide-to-cannabis-testing-regulations.

Royal Commission on Opium. *Final Report.* London: Her Majesty's Stationery Office; 1895.

Rubin V (ed.). *Cannabis and Culture.* The Hague: Mouton; 1975.

Russo EB. Cannabinoids in the management of difficult to treat pain. *Ther Clin Risk Manag.* 2008;4:245–259.

Russo EB. The pharmacological history of cannabis. In: Pertwee R (ed.), *Handbook of Cannabis.* Oxford, UK: Oxford University Press; 2016: 23–43.

Russo EB. Cannabis and epilepsy: An ancient treatment returns to the fore. *Epilepsy Behav.* 2017;70:292–297.

Russo EB, Guy GW, Robson PJ. Cannabis, pain, and sleep: Lessons from therapeutic clinical trials of Sativex, a cannabis-based medicine. *Chem Biodivers.* 2007;4:1729–1743.

Ryberg E, Larsson N, Sjögren S., et al. The orphan receptor GRP55 is a novel cannabinoid receptor. *Br J Pharmacol.* 2007;152:1092–1191.

Satz P, Fletcher JM, Sutker LS. Neurophysiologic, intellectual and personal correlates of chronic marijuana use in native Costa Ricans. *Ann NY Acad Sci.* 1976;282:266–306.

Schicho R, Storr M. A potential role for GBR55 in gastrointestinal functions. *Curr Opin Pharmacol.* 2012;12:653–658.

Schifano F, Orsolini L, Papanti D, Corkery J. Novel psychoactive substances of interest for psychiatry. *World Psychiatry.* 2015;14:15–26.

Schneider M, Schömig E, Leweke FM. Acute and chronic cannabinoid treatment differentially affects recognition memory and social behavior in pubertal and adult rats. *Addict Biol.* 2008;13:345–357.

Schoeler T, Petros N, Di Forti M, et al. Association between continued cannabis use and risk of relapse in first episode psychosis: A quasi-experimental investigation with an observational study. *JAMA Psychiatry.* 2016;73:1173–1179.

Schreiner AM, Dunn ME. Residual effects of cannabis use on neurocognitive performance after prolonged abstinence: A meta analysis. *Exp Clin Psychopharmacol.* 2012;20:420–429.

Schwartz RH, Voth EA, Sheridan MJ. Marijuana to prevent nausea and vomiting in cancer patients: A survey of clinical oncologists. *Southern Med J.* 1997;90:167–172.

Science Daily. Marijuana use disorder is on the rise nationally; Few receive treatment. March 16, 2016. https://www.sciencedaily.com/releases/2016/03/160316105703.htm.

Scuderi C, De Filippis D, Invone T, et al. Cannabidiol in medicine: A review of its therapeutic potential in CNS disorders. *Phytother Res.* 2009;23:597–602.

Serpell MG, Nurmikko TJ, Hoggart B, Toomey PJ, Morln BJ. A multi-centre double blind, randomised, placebo-controlled trial of oromucosal cannabis based medicine in the treatment of neuropathic pain characterised by allodynia [Abstract]. UK Pain Society Meeting, Edinburgh, March 8–11, 2005.

Serpell MG, Nurmikko T, Wright S. Long-term treatment of peripheral neuropathic pain with a phytocannabinoid medicine (Sativex). Paper presented at the British Pharmacology Society Meeting, 2006.

Sewell RA, Poling J, Sofuoglu M. The effect of cannabis compared with alcohol on driving. *Am J Addict.* 2009;18:185–193.

Shang Y, Tang Y. The central cannabinoid receptor type-2 (CB2) and chronic pain. *Int J Neurosci.* 2017;127:812–823.

Shao Z, Yin J, Grzemska M, et al. High resolution crystal structure of the human CB1 cannabinoid receptor. Nature. 2016;540:602–606.

Sherif M, Radhakrishnan R, D'Souza DC, Ranganathan M. Human laboratory studies on cannabinoids and psychosis [Review]. *Biol Psychiatry.* 2016;79:526–538.

Sherva R, Wang Q, Kranzler H, et al. Genome-wide association study of cannabis dependence severity, novel risk variants, and shared genetic risks. *JAMA Psychiatry.* 2016;43:472–480.

Shiplo S, Asbridge M, Leatherdale ST, Hammond D. Medical cannabis use in Canada: Vapourization and modes of delivery. *Harm Reduct J.* 2016;13:30.

Shore DM, Reggio PR. Therapeutic potential of orphan GPCRs, GPR35 and GPR55. *Front Pharmacol.* 2015;6:69–99.

Shrivastava A, Johnston M, Tsuang M. Cannabis use and cognitive dysfunction. *Indian J Psychiatry.* 2011;53:187–191.

Sidney S, Beck JE, Tekawa IS, Quesenberry CP, Friedman GD. Marijuana use and mortality. *Am J Public Health.* 1997;87:585–590.

Silveira MM, Arnold JC, Laviolette SR, et al. Seeing through the smoke: Human and animal studies of 2016. *Neurosci Behav Rev*. 2016;76:380–395.

Smiley A. Marijuana: On road and driving simulator studies. *Alcohol Drugs Driving*. 1986;2:121–134.

Smith AM, Fried PA, Hogan MJ, Cameron I. Effects of prenatal marijuana on visuospatial working memory: An fMRI study in young adults. *Neurotoxicol Teratol*. 2006;28:286–295.

Smith FL, Cichewicz D, Martin ZL, Welch SP. The enhancement of morphine antinociception in mice by delta-9-tetrahydrocannabinol. *Pharmacol Biochem Behav*. 1998;60:559–566.

Smith T, Smith H. Process for preparing cannabine or hemp resin. *Pharm J*. 1846;6:171–173.

Snyder SH. *Uses of Marijuana*. New York: Oxford University Press; 1971.

Solowij N. *Cannabis and Cognitive Functioning*. Cambridge, UK: Cambridge University Press; 1998.

Spaderna M., D'Souza CB. Spicing thing up: Synthetic cannabinoids. *Psychopharmacology (Berl)*. 2011;228:525–540.

Spicaroval D, Nerandzic L, Palecek D. Update on the role of spinal cord TRPV1 receptors in pain modulation. *Physiol. Res*. 2014;63(Suppl. 1): S225–S236.

Steep Hill Cannabis Analysis Laboratory. Quality assurance for medical cannabis. 2010. https://www.steephill.com/pdf/uploads/whitepapers/1c8a41399466e31a99dc38be499bb3a8.pdf.

Steffens S, Vellard NR, Arnaud C, et al. Low dose oral cannabinoid therapy reduces progression of atherosclerosis in mice. *Nature*. 2005;2434 :782–786.

Steinmetz K. What marijuana legalization in Canada could mean for the United States. *TIME*. April 6, 2017.

Substance Abuse and Mental Health Services Administration. US National Survey on Drug Use, 2015 (detailed tables, SAMHSA, CBHSQ). 2015. http://www.samhsa.gov/data/sites/default/files/NSDUH-DetTabs-2015/NSDUH-DetTabs-2015/NSDUH-DetTabs.

Substance Abuse and Mental Health Services Administration. National Survey on Drug Use and Health. 2016. https://www.samhsa.gov/data/population-data-nsduh.

Sullan SR, Millar SA, England TJ, O'Sullivan SE. A systematic review and meta-analysis of the haemodynamic effects of cannabidiol. *Front Pharmacol*. 2017;70:328–333.

Swift A. Support for legal marijuana use up to 60% in U.S. Gallup, Social Issues. October 19, 2016. http://www.gallup.com/poll/196550/support-legal-marijuana.aspx.

Swift W, Hall W, Teesson M. Cannabis use and dependence among Australian adults: Results from the National Survey of Mental Health and Wellbeing. *Addiction*. 2001;96:737–748.

Syad YY, McKeage K, Scott LJ. Delta-9-terahydrocannabinol (Sativex): A review of its use in patients with moderate to severe spasticity due to multiple sclerosis. *Drugs*. 2014;74:563–578.

Szabo B. Effects of phytocannabinoids on neurotransmission in the central and peripheral nervous system. In: Pertwee R (ed.), *Handbook of Cannabis*. Oxford, UK: Oxford University Press; 2015: 157–172.

Szabo B, Schlicker E. Effects of cannabinoids on neurotransmission. *Handb Exp Pharmacol*. 2005;168:328–365.

Tabrizi M, Baraldi PG, Baraldi S, et al. Medicinal chemistry, pharmacology and clinical implications of TRPV1 receptor antagonists. *Med Res Rev*. 2017;37:936–983.

Talking Drugs. Interactive map: Medical cannabis in the European Union. Released August 8, 2017. https://www.talkingdrugs.org/interactive-map-medical-cannabis-in-european-union.

Tanda G, Munzar P, Goldberg GR. Self-administration behavior is maintained by the psychoactive ingredient of marijuana in squirrel monkeys. *Nat Neurosci*. 2000;3:1073–1074.

Tanda G, Pontieri FE, Di Chiara G. Cannabinoid and heroin activation of mesolimbic dopamine transmission by a common μ_1 opioid receptor mechanism. *Science*. 1997;276:2048–2050.

Tashkin DP. Smoked marijuana as a cause of lung injury. *Monalid Arch Chest Dis*. 2005;63:93–100.

Tashkin DP. Effects of marijuana smoking on the lung. *Ann Am Thorac Soc*. 2013;10:239–247.

Tashkin DP, Coulson AH, Clark VA, et al. Respiratory symptoms and lung function in habitual heavy smokers of marijuana alone, smokers of marijuana and tobacco, smokers of tobacco alone, and nonsmokers. *Annu Rev Respir Dis*. 1987;135:209–216.

Tashkin DP, Simmons MS, Sherrill DL, Coulson AH. Heavy habitual marijuana smoking does not cause an accelerated decline in FEV1 with age. *Am J Respir Crit Care Med*. 1997;155:141–148.

Thakur GA, Nikas SP, Li C, Makryannis A. Structural requirements for cannabinoid receptor probes. *Handb Exp Pharmacol*. 2005;168:209–246.

Thirthalli J, Benegal V. Psychosis among substance users. *Curr Opin Psychiatry*. 2006;19:239–245.

Thomas H. Psychiatric symptoms in cannabis users. *Br J Psychiatry*. 1993;163:141–149.

Todd AR. Hashish. *Experientia.* 1946;2:55–60.

Torbjörn U, Järbe C, Henriksson BG. Discriminative response control produced with hashish, tetrahydrocannabinols, and other drugs. *Psychopharmacology.* 1974;40:1–16.

Tramer MR, Carroll D, Campbell FA, et al. Cannabinoids for control of chemotherapy induced nausea and vomiting: Quantitative systematic review. *Br Med J.* 2001;323:16–21.

Transform. Drug decriminalisation in Portugal: Setting the record straight. June 11, 2014. https://www.tdpf.org.uk/blog/drug-decriminalisation-portugal-setting-record-straight.

Turner SE, Williams CM, Iversen L, Whalley BJ. Molecular pharmacology of phytocannabinoids. *Prog Chem Nat Prod.* 2017;103:61–101.

United Nations Office on Drugs and Crime. The UN Conventions 1961 and 1971. 2017. https://www.unodc.org/unodc/en/commissions/CND/Mandate_Functions/Mandate-and-Functions_Scheduling.html.

Valverde O, Karsak M, Zimmer A. Analysis of the endocannabinoid system by using CB_1 cannabinoid receptor knockout mice. *Handb Exp Pharmacol.* 2005;168:117–146.

Van Amsterdam JGC, van der Laan JW, Slangen JL. Residual effects of prolonged heavy cannabis use. Report No. 318902003. Bilthoven, the Netherlands: National Institute of Public Health and the Environment; 1996.

Van Sickle MD, Duncan M, Kingsley PJ, et al. Identification and functional characterization of brainstem cannabinoid CB_2 receptors. *Science.* 2005;310:329–332.

Varma N, Carlson GC, Ledent C, Alger BE. Metabotropic glutamate receptors drive the endocannabinoid system in hippocampus. *J Neurosci.* 2001;21:RC188.

Vaughan VW, Christie MJ. Retrograde signalling by endocannabinoids. *Handb Exp Pharmacol.* 2005;168:368–383.

Veksler K. Canada announces plans to legalize cannabis in 2018. *Mass Roots.* March 30, 2017. https://www.massroots.com/news/is-canada-legalizing-cannabis.

Velasco G, Sanchez C, Guzman M. Endocannabinoids and cancer. *Handb Exp Pharmacol.* 2015;231:449–472.

Vermersch P. Sativex (tetrahydrocannabinol + cannabidiol), an endocannabinoid modulator: Basic features and main clinical data. *Expert Rev Neurother.* 2011;4(Suppl.):15–19.

Volkow ND, Swanson NM, Evins AE, et al. Effects of cannabis use on human behavior, including cognition, motivation, and psychosis: A review. *JAMA Psychiatry.* 2015;73:292–297.

Wade D, Collin C, Stephens M. Meta-analysis of the effects of Sativex on spasticity in MS subjects. JMS Clin Lab Res. 2005;11(Suppl 1):S97.

Wade DT, Makela PM, House H, Bateman C, Robson P. Long-term use of a cannabis-based medicine in the treatment of spasticity and other symptoms in multiple sclerosis. *Multiple Sclerosis.* 2006;12:639–645.

Walker JM, Hohmann AG. Cannabinoid mechanisms of pain suppression. *Handb Exp Pharmacol.* 2005;168:511–544.

Walton RP. Marihuana: America's New Drug Problem. New York: Lippincott; 1938.

Wang H, Dey SK, Maccarrone M. Jekyll and Hyde: Two faces of cannabinoid signalling in male and female fertility. *Endocrin Rev.* 2006;27:427–448.

Warburton H, May T, Hough M. Looking the other way: The impact of reclassifying cannabis on police warnings, arrests and informal action in England and Wales. *Br J Criminol.* 2005;45:113–128.

Ward SJ, Raffa RB. Rimonbant redux and strategies to improve the future outlook of CB1 receptor neutral-antagonist/inverse-agonist therapies. *Obesity.* 2011;19:1325–1334.

Ware MA, Adams H, Guy GW. The medicinal use of cannabis in the UK: Results of a nationwide survey. *Int J Clin Pract.* 2005;59:291–295.

Warholak M. How cannabis affects fertility and pregnancy. Health MJ, June 3, 2016. http://www.healthmj.com/cancer/cannabis-affects-fertility-pregnancy.

Wealth Daily. 3 Legal marijuana stocks to own now. August 5, 2017. http://secure.wealthdaily.com/96292?device=c&gclid=EAIaIQobChMIx4L3zea_1QIV1RXTCh0kyw4KEAAYASAAEgK5RfD_BwE.

Weil AT, Zinberg NE, Nelsen JM. Clinical and psychological effects of marihuana in man. *Science.* 1968;162:1234–1242.

Weiland BJ, Depue BE, Sabbineni A. Daily marijuana use is not associated with brain morphometric measures in adolescents or adults. i. 2015;35:1505–1512.

Whiting PF, Wolff RF, Deshpande S, et al. Cannabinoids for medical use: A systematic review and meta-analysis. *JAMA.* 2015;313:2456–2473.

Wiley JL, Barret RL, Lowe J, Balster RL, Martin BR. Discriminative stimulus effects of CP-55,940 and structurally dissimilar cannabinoids in rats. *Neuropharmacology.* 1995;34:669–676.

Wiley JL, Marusich JA, Huffman JW. Moving around the molecule: Relationship between chemical structure and in vivo activity of synthetic cannabinoids. *Life Sci.* 2014;97:55–63.

Wiley JL, Owens RA, Lichtman JH. Discriminative stimulus properties of phytocannabinoids, endocannabinoids, and synthetic cannabinoids. *Curr Top Behav Neurosci.* June 9, 2016 [Epub ahead of print].

Williams E. Three G Farms cleared to begin planting marijuana: Colfax-area operation will raise and process pot. *Lewiston Tribune,* February 13, 2015. http://lmtribune.com/northwest/three-g-farms-cleared-to-begin-planting-marijuana/article_c10b8f50-c86b-5915-80a3-46cd73f8f15e.html.

Williams JH, Wellman NA, Rawlins JNP. Cannabis use correlates with schizotypy in healthy people. *Addiction.* 1996;91:869–887.

Wilson RI, Nicoll RA. Endogenous cannabinoids mediate retrograde signalling at hippocampal synapses. *Nature.* 2001;410:588–592.

Wirguin I, Mechoulam R, Breuer A, et al. Suppression of experimental autoimmune encephalomyelitis by cannabinoids. *Immunopharmacology.* 1999;28:209–214.

Wisset R. *A Treatise on Hemp.* London: Harding; 1808.

Witkin JM, Tzavara ET, Davis RJ, Li X, Nomikos G.G. A therapeutic role for cannabinoid CB-1 receptor antagonists in major depressive disorders. *Trends Pharmacol. Sci.* 2005;26:609–617.

Wood GB, Bache F. *The Dispensatory of the United States.* Philadelphia, PA: Lippincott; 1854: 339.

Woodhams SG, Sagar DR, Burston JJ, Chapman V. The role of the endocannabinoid system in pain. *Handb Exp Pharmacol.* 2015;227:119–143.

Wu TC, Tashkin DP, Djaheb B, Rose JE. Pulmonary hazards of smoking marijuana as compared with tobacco. *N Engl J Med.* 1988;31:347–351.

Yankey BA, Rothenberg R, Strasser S, et al. Effect of marijuana use on cardiovascular and cerebrovascular mortality: A study using the National Health and Nutrition Examination Survey linked mortality file. *Eur J Prev Cardiol.* 2017;24:1833–1840.

Zajicek J, Fox P, Sanders H, et al. Cannabinoids for treatment of spasticity and other symptoms related to multiple sclerosis (CAMS study): Multicentre randomised placebo-controlled trial. *Lancet.* 2003;362:1517–1526.

Zalesky A, Solowij N, Yücel M, et al. Effect of long-term cannabis use on axonal fibre connectivity. *Brain.* 2012;135:2245–2255.

Zammit S, Allebeck P, Andreasson S, Lundberg I, Lewis G. Self reported cannabis use as a risk factor for schizophrenia in Swedish conscripts of 1969: Historical cohort study. *Br Med J.* 2002;325:1195–1212.

Zhang LR, Morgenstern H, Greenland S, et al. Cannabis smoking and lung cancer risk: Pooled analysis in the International Lung Cancer Consortium. *Int J Cancer.* 2015;136:894–903.

Zimmer A, Zimmer AM, Hohmann AG, Herkenham M, Bonner TI. Increased mortality, hypoactivity, and hypoalgesia in cannabinoid CB1 receptor knockout mice. *Proc Natl Acad Sci USA.* 1999;96:5780–5785.

Zimmerman S, Zimmerman A.M. Genetic effects of marijuana. *Int J Addiction.* 1990;25:19–33.

Ziring D, Wei B, Velazquez P, Schrage M, Buckley NE, Braun J. Formation of B and T cell subsets require the cannabinoid receptor CB2. *Immunogenetics.* 2006;58:714–725.

Zogopoulos P, Vasileiou I, Patsouris E, Theocharis SE. The role of endocannabinoids in pain modulation. *Fundam Clin Pharmacol.* 2013;27:64–80.

INDEX